Machine Vision

Japanese Technology Reviews

Japanese Technology Reviews

Manufacturing Engineering

Automobile Electronics
Shoichi Washino

Steel Industry I: Manufacturing System
Tadao Kawaguchi and Kenji Sugiyama

Steel Industry II: Control System
Tadao Kawaguchi and Takatsugu Ueyama

Networking in Japanese Factory Automation
Koichi Kishimoto, Hiroshi Tanaka,
Yoshio Sashida and Yasuhisa Shiobara

Biotechnology

Production of Nucleotides and Nucleosides by Fermentation
Sadao Teshiba and Akira Furuya

Recent Progress in Microbial Production of Amino Acids
Hitoshi Enei, Kenzo Yokozeki and Kunihiko Akashi

Electronics

MMIC—Monolithic Microwave Integrated Circuits
Yasuo Mitsui

Bulk Crystal Growth Technology
Shin-ichi Akai, Keiichiro Fujita, Masamichi Yokogawa, Mikio
Morioka and Kazuhisa Matsumoto

Semiconductor Heterostructure Devices
Masayuki Abe and Naoki Yokoyama

Computers and Communications

**Machine Vision: A Practical Technology for Advanced Image
Processing**
Masakazu Ejiri

This book is part of a series. The publisher will accept continuation orders which may be cancelled at any time and which provide for automatic billing and shipping of each title in the series upon publication. Please write for details.

Machine Vision
A Practical Technology
for Advanced Image Processing

by
Masakazu Ejiri
Hitachi, Ltd.
Tokyo, Japan

Gordon and Breach Science Publishers

New York • London • Paris • Montreux • Tokyo • Melbourne

Gordon and Breach Science Publishers

Post Office Box 786
Cooper Station
New York, New York 10276
United States of America

Post Office Box 197
London WC2E 9PX
United Kingdom

58, rue Lhomond
75005 Paris
France

Post Office Box 161
1820 Montreux 2
Switzerland

3-14-9, Okubo
Shinjuku-ku, Tokyo 169
Japan

Private Bag 8
Camberwell, Victoria 3124
Australia

Portions of this work were originally published in Japanese as (Kogyoyo guzoshori [Industrial Image Processing]), © 1988 by Shokodo Co. Ltd., Tokyo. Reprinted by permission of Shokodo Co. Ltd., Tokyo.

Library of Congress Cataloging-in-Publication Data

Ejiri, Masakazu.
  Machine vision / by Masakazu Ejiri.
    p. cm.—(Japanese technology reviews ; v. 10)
  Bibliography: p.
  Includes index.
  ISBN 2-88124-353-3
  1. Computer vision. I. Title. II. Series.
TA1632.E35   1989   89-31228
621.39′9—dc20   CIP

Contents

Preface to the Series

Modern technology has a great impact on both industry and society. New technology is first created by pioneering work in science. Eventually, a major industry is born, and it grows to have an impact on society in general. International cooperation in science and technology is necessary and desirable as a matter of public policy. As development progresses, international cooperation changes to international competition, and competition further accelerates technological progress.

Japan is in a very competitive position relative to other developed countries in many high technology fields. In some fields, Japan is in a leading position; for example, manufacturing technology and microelectronics, especially semiconductor LSIs and optoelectronic devices. Japanese industries lead in the application of new materials such as composites and fine ceramics, although many of these new materials were first developed in the United States and Europe. The United States, Europe, and Japan are working intensively, both competitively and cooperatively, on the research and development of high-critical-temperature superconductors. Computers and communications are now a combined field that plays a key role in the present and future of human society. In the next century, biotechnology will grow, and it may become a major segment of industry. While Japan does not play a major role in all areas of biotechnology, in some areas such as fermentation (the traditional technology for making "sake"), Japanese research is of primary importance.

Today, tracking Japanese progress in high-technology areas is both a necessary and rewarding process. Japanese academic institutions are very active; consequently, their results are published in scientific and technical journals and are presented at numerous meetings where more than 20,000 technical papers are presented orally every year. However, due principally to the language barrier, the results of academic research in Japan are not well known overseas. Many in the United States and in Europe are thus surprised by the sudden appearance of Japanese high-technology products. The products are admired and enjoyed, but some are astonished at how suddenly these products appear.

With the series *Japanese Technology Reviews,* we present state-of-the-art Japanese technology in five fields:

> Electronics,
>
> Computers and Communications,
>
> Manufacturing Engineering,
>
> New Materials, and
>
> Biotechnology.

Each tract deals with one topic within each of these five fields and reviews both the present status and future prospects of the technology, mainly as seen from the Japanese perspective. Each author is an outstanding scientist or engineer actively engaged in relevant research and development.

We are confident that this series will not only give a bright and deep insight into Japanese technology but will also be useful for developing new technology of our readers' own concern.

As editor in chief, I would like to acknowledge with sincere thanks the members of the editorial board and the authors for their contributions to this series.

TOSHIAKI IKOMA

Preface

Human beings have an important faculty, called vision, with which they can instantly identify objects, their positions and situations, by looking at the outside world with their eyes. If this faculty could be realized by engineering means, many machines and much equipment could be furnished with the same capability. Machines with vision could then perceive changes to their environmental situations and modify their behavior to adapt to these changes. Thus, the machines could be made highly flexible and even intelligent, and, consequently, be effective in performing useful tasks on behalf of human beings, especially in many industrial applications. Image processing, among others, is a key technology for realizing machine vision. This is obvious because human beings also process images projected onto the retina.

Image processing covers a variety of areas, including image acquisition, image transformation, image enhancement, image compression, image restoration, image generation, image transmission, image accumulation and storage, and image recognition and understanding. Objects to be processed also vary in form, including documents consisting mainly of characters, drawings consisting mainly of figures, pictures comprising gray or color information, and scenes containing three-dimensional objects. However, in industrial applications, analysis of scenes containing three-dimensional objects is especially important. Thus, recognition and understanding of objects are central to machine-vision research. Many distinctive image-processing techniques are required for machine vision arising from the three-dimensional feature of the objects, in addition to techniques common to all image-processing areas.

In machine vision, high-speed image-processing capability in a real-time mode is one of the key issues for industrial applications such as the automation of production lines. Therefore, methodology-oriented basic research, sometimes called computer vision, as well as practical research toward realization of dedicated processing hardware, including image-processing LSIs, have been at the core of robotics research for the twenty-five years of its history. Japanese industry took the initiative in creating this new technological field, and has since been playing a major role in its advancement by developing many useful concepts and world-leading machine-vision systems. The author has also been involved deeply in this technological field since its inception.

Machine-vision technology, however, is not a well-established technology. We are only witnessing its beginnings. Still very little is known about human vision, especially how human beings recognize three-dimensional objects. The technological level so far attained is very rudimentary compared to that of human vision. Methods developed so far are obviously very different from those used by human vision. Therefore, this technological field still presents many challenges, and many problems remain to be solved. The exploitation of various new applications may be the key to further advancement of both basic and applied technologies for future machine-vision systems.

This tract has been planned to present the state of the art of machine vision, especially focusing on the practical technologies developed in Japan. Some introductory remarks are made in Chapter 1 and a few basic and useful algorithms are explained in Chapter 2, together with their processing circuits, which are widely applicable in industrial machine-vision systems. In Chapters 3 through 5, methods and machines that recognize three types of basic features of objects are introduced; these are geometric features, such as shape and position; qualitative features, such as flaws and scratches; and supplemental features, such as marks and symbols on the object surfaces. In Chapter 6, image processors that have been developed and are being used in industry are summarized. Some important aspects of machine vision are also discussed together with a brief discourse on future perspectives. This tract is written from the applications-oriented standpoint, so that readers such as researchers, engineers, and college students can easily learn practical image-processing technologies that are useful in realizing machine vision. Nothing will make the author happier than if he can stimulate younger readers to enter this field.

In pursuing the study of machine vision and in preparing this tract, the author has received much encouragement and many suggestions and comments. His special thanks go to Prof. Toshiaki Ikoma and Dr. Kazumoto Iinuma, editors of the series *Japanese Technology Reviews*. Those whose names follow are also singled out for special appreciation: Yasutsugu Takeda, Toshio Numakura, Jun Kawasaki, Takeshi Uno, Sadahiro Ikeda, Seiji Kashioka, Haruo Yoda, Hitoshi Matsushima, Michihiro Mese, Toshikazu Yasue, Hirotada Ueda, Yoshihiro Shima, Hiroshi Sakou, Jun Motoike, and Takafumi Miyatake.

This tract is based partly on the author's previous book, *Kogyoyo gazoshori* (*Industrial Image Processing*), published in Japanese by the Shokodo Co. Ltd., Tokyo (1988). The author is particularly grateful to Mr. Takao Kobayashi of Shokodo for extending every convenience in publishing this new English tract. The author also thanks Miss Miki Kasuya for her excellent work in preparing the draft.

MASAKAZU EJIRI

CHAPTER 1

Introduction to Machine Vision

1.1. Vision and Robotics

1.1.1. History in Brief

All observable objects existing in the physical world possess their own three-dimensional (3D) shapes, except for characters, symbols, and figures created by human intelligence. To enable a computer to recognize these 3D objects has been a subject of artificial-intelligence research started in the late 1960s as a part of research into intelligent robots. This research has been called "object recognition" to distinguish it from other recognition problems for purely two-dimensional (2D) objects, such as character-recognition problems that were prevalent at the time. This has also been called generically "computer vision" because it is an effort to add a visual function to a computer. In the early stages of research, blocks were chosen as objects to be recognized because of the simplicity in describing their shapes within the computer. Analysis of more complex objects, such as desktop scenes, indoor scenes, and outdoor scenes, became of increasing interest. This research has been aimed not only at recognition of objects in a scene, but also at understanding the whole scene by describing the mutual relations between objects. In this sense, the research has been called "scene analysis" and has resulted in many unique methodologies.

In Japan, research in scene analysis also started in the late 1960s. However, the objectives and approaches were somewhat different from those in other parts of the world. In addition to basic research on recognition methodology, much stress was placed on practical research. This resulted in successive development of many unique and useful image processors for industrial applications, featuring the ability to analyze inputting images in a real-time mode. Thus, research on vision technology, together with research on industrial robots, formed the basis of a new research field called "robotics," which began in the mid-1970s. This led to innovations in the manufacturing technology area, with progress in semiconductor-device technology typified by microcomputers and memories. In particular, the contribution of vision technology to the assembly of semiconductor devices was enormous. These innovations led to the successful automation of various other production lines, enabling labor savings, productivity improvement, and product quality improvement. In recent years, much research has been conducted to further stimulate the application of vision technology. The term "machine vision" is becoming well known, stressing the application-oriented aspect more than in computer vision. In this book, machine-vision technology is extensively dealt with, drawing the reader's attention to practical productive vision applications.

1.1.2. Intelligent Machines

Attempts to make a machine intelligent by adding vision and other intelligent functions will become increasingly important in various future industrial applications. Such an intelligent machine can generally be represented by the configuration shown in Figure 1. Functions needed for an intelligent machine are summarized as follows:

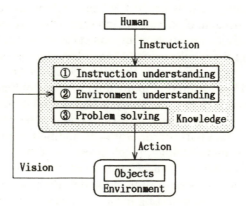

Figure 1. An intelligent machine.

1. Understanding instructions.
2. Understanding environments.
3. Solving problems.

When a human gives an instruction to a conventional machine, the instruction is liable to be very microscopic, having a one-to-one correspondence with each motion of the machine. This means that the human is initiating each consecutive movement one by one, to achieve a certain objective or task, by resolving it into a sequence or action plan. In other words, the human is teaching the machine "how to behave." Thus, the instruction is usually very detailed, and giving the instruction to the machine is a tedious job for the human operator. If the machine is sufficiently intelligent, it may be able to understand a set of operations from a simple macroscopic instruction given by the human operator. It may even be possible for the machine to understand the objective from the macroscopic instruction and to resolve the objective by choosing the appropriate sequence of operations from its own knowledge. In this case, the macroscopic instruction must preferably involve "what-to-achieve" information instead of "how-to-behave" information.

There are two means in principle of giving an intelligent machine a macroscopic instruction.[1] These are

1. Linguistic means.
2. Pictorial means.

The first means includes methods based on higher-level languages, and is based on natural languages as the ultimate goal. The second means includes showing assembly drawings and circuit diagrams and having the machine interpret them.[2,3] Thus, instruction understanding varies from microscopic to macroscopic and from "how" to "what." It is becoming one of the key issues of the intelligent man–machine interface, and will form an important field in future artificial intelligence research.

Understanding of the environment is the facility of investigating the outside situation by using vision (and other sensing means in some cases) and of recognizing the objects and their relations, thereby understanding the present state of the environment in which the machine itself is involved. Thus, this facility can provide the starting condition for machine behavior.

The problem-solving facility usually takes account of the starting condition provided by the environment-understanding function and of the goal provided by the instruction-understanding function, and plans how to behave from the present condition to the goal. Among the plans conceived in this task-planning process, a decision is made to select the most appropriate plan for execution. If an "unexpected situation" occurs during the action, the environment-understanding function is kept in continuous operation, and every time it finds a change, the plan is modified in real time to adapt to the new situation.

1.1.3. Image Processing

As stated before, vision is an indispensable function for intelligent machines, and their core technology is image processing. In general, image processing involves several factors: converting an image into another more important image, converting an image into nonimage data, and converting nonimage data into an image. Conversion of images into nonimage data is called analytical image processing, and conversion of nonimage data to images is called synthetic image processing. A typical example of the latter is computer tomography.

Pattern recognition is an analytic image-processing method that outputs the name of the object involved in the input image. For example, outputting a character name from a character pattern is called character recognition, and outputting the name of a person from a portrait can be called personal recognition or, more preferably, personal identification. Pattern recognition to output object names by looking at the three-dimensional real world may simply be called object recognition.

Another kind of analytic image-processing method, called image understanding, outputs a description of the image. This aims at describing what types of objects are included in the image and in what kind of relationships. That is, image understanding is the function of recognizing both the objects and their mutual relationships. When the image is a projection of a three-dimensional view of the outside world, image understanding can be called scene analysis. The resulting output is generally a structural description of the scene. An example of such a description is "two blocks stand separately at a certain distance and another block lies on top of them, thus forming an arch." This type of description is usually made by using symbolic logic such as predicate calculus. To make the machine understand an image like this, it is necessary to furnish the machine with several types of knowledge.

Object recognition and scene analysis, two of many types of image processing, are very important as key technologies in robotics-related industrial-vision systems.

1.2. Scene Analysis

1.2.1. Passive Method

The passive method images a scene as it is, analyzes it, and recognizes objects and their relationships, without applying any specifically structured illuminations. There are two basic procedures in the passive method, as shown in Figure 2. One is based on edge extraction and the other is based on plane extraction. Edge extraction utilizes a spatial differentiation of the input image to extract edge information of objects. Detected edge lines are combined to form planes. On the other hand, plane extraction combines pixels

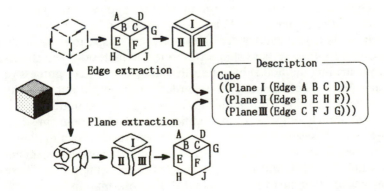

Figure 2. Two procedures in the passive method.

with similar brightness and/or textural features into surface elements. These surface elements are combined with the adjacent resembling surface elements, thus producing regional clusters. The boundaries of these clusters, where further merging is impossible, are finally regarded as edge lines. Thus, an object is represented as a list of points, lines, and planes, forming a description of an object. This object is then compared with the object models stored in the machine and identified.

To obtain truly three-dimensional information of the object from a single view of a monocular vision system, the machine must be furnished with knowledge of the object *a priori*. Though binocular vision is also a type of passive method, it can obtain three-dimensional information without using any particular knowledge of the object. In this method, corresponding points between two views obtained from left and right imaging devices are found by using regional correlation. Disparity between the two views is utilized to calculate the distance from the imaging devices to the object, based on triangulation. Thus, the images represented at first as a set of brightness data can be converted into a set of range data. This set is called a distance map, or depth map, and each pixel value represents the distance to a corresponding point on the object. For example, the larger pixel values on the distance map represent the nearer portions of the object, thus making it possible to extract the nearer objects by simply thresholding the distance map. Thus, the distance map can be used to provide effective information for separating objects and estimating their shapes.

There are also a number of methods for shape estimation, generally called shape-from-X methods, in which the character X represents a particular word. Examples are shape-from-shading, shape-from-texture, and shape-from-motion methods. The shape-from-shading method is based on the analysis of the brightness of each pixel. The brightness is regarded as an important quantity resulting from light reflection on an object surface under known illumination conditions. One modification of the shape-from-shading method is called photometric stereo, where a number of illuminations from known directions are switched on one by one to obtain a number of different images from a single viewing point. These images are then analyzed to yield distance information.

1.2.2. Active Method

The active method projects waves to the objects and extracts distance information from observation of their reflections. The method actively utilizes waves as media for conveying distance information. A laser light can be used as a range finder for point-to-point measurement in the outside world, based on either time of flight between emission and reflection or phase shift between modulated light emission and reflection.

Other conventional active methods are shown in Figure 3, in which a structured light such as a spotlight, a slit light, a grid light, or even a patterned light is projected. An imaging device such as a TV camera is set apart from the light-projection point, and observes the image of the projected light. The position of the observed light in the image is directly converted to the three-dimensional position of the corresponding point on the object based on triangulation.

The spotlight method is effective for quickly defining the outline of objects by following their contours. Therefore, it can be utilized conveniently in simple robot vision for avoiding obstacles. The slit-light method observes a cross section of an object as if it were cut

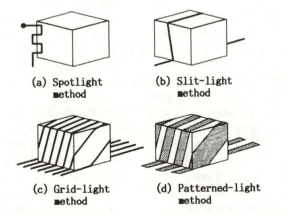

(a) Spotlight method

(b) Slit-light method

(c) Grid-light method

(d) Patterned-light method

Figure 3. Active methods using structured light: (a) spotlight method, (b) slit-light method, (c) grid-light method, and (d) patterned-light method.

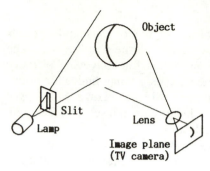

Figure 4. Principle of slit-light method.

by a light plane, as shown in Figure 4. An example of an image obtained by the slit-light projection method is shown in Figure 5, where Figure 5(a) is the combined image observed for incremental positions of the slit light, and Figure 5(b) is the resulting three-dimensional direction of each surface element calculated from the distance information at each position.[4] As bright light can be used in the active method, stable images are usually obtained, thus ensuring high reliability.

The grid-light method is utilized with Moire analysis or Fourier analysis. The Moire method observes the projected grid through another grid, thus forming a Moire fringe pattern that is an equidistant contour map of the object. In the projected grid image, direction changes and density changes of the grid lines are observed for each object plane having different surface normals. Therefore, it is possible to separate each plane surface of objects such as polyhedra by transforming the image into a frequency region using a two-dimensional Fourier transform.[5]

One type of patterned-light method projects a coded light pattern to the three-dimensional world, thereby supplying different codes to each sectioned region of the world. A typical method utilizes a Gray-coded light-pattern generator and projector, controlled by a liquid-crystal switch.[6] As shown in Figure 6, the pattern generator sequentially projects n types of Gray-code pattern, thereby discriminating the three-dimensional space into 2^n striped regions when viewed from the direction in which light is projected. Each region is thus designated with a different n-bit code. When

(a) Slit-light image

(b) Surface normals

Figure 5. Slit-light method and resulting surface normals: (a) slit-light image and (b) surface normals. (From Ohshima, Reference 4.)

the brightness at a certain pixel position in the TV image is observed, a sequence of n values is obtained, each with logical 0 or 1. This pattern sequence corresponds to a certain spatial code. Therefore, the direction of illumination is observed for each pixel. As the viewing direction of the pixel is calculated from the pixel position for each pixel, the three-dimensional position of the point (P) on the object surface can be calculated.

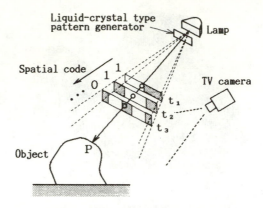

Figure 6. Patterned-light method.

Another patterned-light method utilizes a specific texture pattern that is known *a priori*. The projected texture image is then observed, and the deformation of each texture pattern and/or density changes between each texture pattern is calculated to find the slope of the planes. This method applies a procedure generally called texture analysis. Textures consisting of simpler patterns are preferred for ease of measurement. These texture patterns include circular and diamond-shaped patterns. This method simplifies recognition of plain objects with no texture by adding a texture by illumination.

1.3. Industrial Vision

1.3.1. Object Features

The objects to be recognized have, in general, three types of features, as shown in Figure 7. One is called geometric and relates to the shape, size, type, position, and orientation of the object. These are features involving linear measurement. They can be considered by classifying them into two categories: one, shape features, including size and type of objects, and the other, positional features,

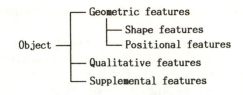

Figure 7. Object features.

including orientation. Recognition of shape features is important for automating classification and selection in the production process, whereas recognition of positional features is important specifically for automation of assembly processes, although it is also important for every aspect of industrial vision.

Objects sometimes have flaws such as cracks, scratches, and faded color. They may also give an impression of softness, coldness, etc. to the human observer. It is difficult to make a machine recognize these features or feel these impressions. These features are of the second type, and are called qualitative features. Visual inspection of these qualitative features is carried out by human inspectors in a fairly large number of steps in the production process. To automate these processes, qualitative features must also be represented quantitatively by converting them into measurable parameters. Only those features that are convertible to quantitative features can be successfully automated.

The third type, called supplemental features, involves marks, characters, textures, and colors that are added to object surfaces. As an example, reading of a printed product name on an object surface is very important in selection and classification of objects in automated physical-distribution systems.

Of these three types of features, recognition of geometric features such as shape and position in particular requires handling of three-dimensional information. Therefore, geometric-feature recognition is of great concern to many researchers in theoretical computer vision. However, it is also becoming increasingly important, especially for machine-vision practitioners, to find effective methods for recognizing qualitative and supplemental features in a practical sense.

1.3.2. Reduction into a Two-dimensional Problem

Though a variety of scene-analysis techniques has been studied in many places around the world, no single method has been found to be applicable to every aspect of object recognition. The functions of these techniques are still far inferior to those of human-vision systems. Processing speed and recognition reliability are particular problems. For example, methods utilizing a number of images, in particular those based on search procedures for corresponding points between images, are slow. Photometric stereo method uses the switching of light sources, which sometimes poses undesirable problems in industrial applications. This is because the light source requires careful positional setting at initial installation, frequent lamp changes due to limited lamp life, and brightness stabilization each time the light source is switched from one to another. Settling time at each switching is an extra overhead for recognition.

In most scene-analysis research, objects are observed from an oblique direction to obtain a generalized method. However, this is unnecessary in the application of machine vision, in particular for recognizing objects moving on a conveyor. It is also unnecessary to describe the whole object and to completely understand the whole scene. There are many cases where only a single plan view is enough for analysis because the types of objects to be recognized are usually limited in number in a given industrial application, and the shapes are known *a priori* and do not change during the recognition process. When the objects are tumbled on a table, they usually settle to one of a limited number of stable configurations. Therefore, it is possible to recognize an object from a single view, if the possible plan images or their analyzed results are memorized in advance for comparison purposes. An early attempt along these lines was the rotational pattern-matching method,[7] as illustrated in Figure 8. In the figure, the top image is converted into a polar-coordinate pattern around the center of gravity of the object. In the recognition process, the input and the preregistered standard patterns are compared by shifting the angular axis.

Some cases occur in industrial applications where objects are intrinsically planar and only the top views can be taken into account, such as in the inspection of printed-circuit boards and IC wafers.

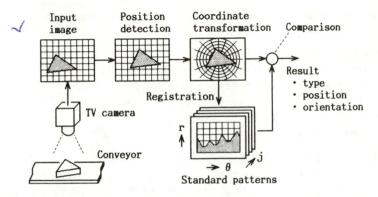

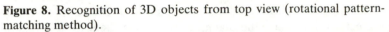

Figure 8. Recognition of 3D objects from top view (rotational pattern-matching method).

Even if an object has a complex 3D shape, one surface of the object can be positively constrained to a standard plane when it is fed into the field of view of an imaging device, thus reducing the 3D problem to a 2D problem. Thus, the object can be analyzed with minimal essential operations, using the least number of images, possibly just one. Restricting the fed-in configuration into one of a limited number of possible configurations usually minimizes system cost. Furthermore, the image is, in most cases, thresholded into a binary form at an early processing stage, so that subsequent processing can be executed by logical operations, thus minimizing hardware scale. To execute such processes, a specially designed image processor is usually used, as a conventional general-purpose computer alone does not meet the processing speed requirements. Recent progress in LSI technology facilitates the realization of general-purpose image processors, into which an application-oriented algorithm is easily implemented. However, there are many other problems to be solved by applying specially designed image processors to minimize the cost/performance ratio.

1.3.3. Requirements for Vision

For living organisms, vision is essential for recognizing the outside world. For automatic industrial machines, it is also essential for sensing the surrounding environment. The basic functions needed for industrial vision systems are to judge

- whether or not an object exists,
- whether or not the object is in its proper position,
- whether or not the object is in its correct orientation,
- whether or not the object is a desirable object,
- whether or not the object is normal, and
- what the object is.

If the machine can judge these matters through vision, it can be made to modify its behavior by responding to its judgment. For example, if an undesirable object is fed in to be processed, the machine can remove it. If the position and orientation of a fed-in object are incorrect, the machine can adjust them for subsequent assembly operations. Thus, we can expect improved efficiency, and improved product quality. Vision is thus a very important means for automating production lines.

There are three basic requirements for industrial-vision systems. One is processing speed, which usually ranges from 0.1 to 1 second for outputting the judged result after receiving the object into the field of view of an imaging device. One way of attaining this speed is to utilize an image processor between the imaging device and the computer. Another way is to use a multiprocessor computer consisting of parallel processing units.

The second requirement is reliability of judgment, which typically requires a false-alarm rate and error rate of less than 0.1%. The most important factor for this is stable imaging. The distance between imaging device and object should be kept as constant as possible. Stabilization of the illumination power source is also important. Another important factor for reliability is a simple and robust algorithm. The means of absorbing time variations of absolute and relative brightness (contrast) is essential for robustness. Sometimes, redesigning the object and rearranging the background in which the object is placed may help simplify the recognition algorithm. These factors are important in realizing a reliable vision system that is applicable in industry.

The third requirement is cost. This is sometimes the most critical requirement because the budget is often insufficient. The development cost of an industrial machine-vision system should, of course, be matched to the advantages to be gained by its introduction. Therefore, the algorithm may sometimes have to be replaced by a less costly one, even though it may be the most effective for a given purpose.

1.3.4. History of Industrial Vision

Image processing, which is a key function for realizing machine vision, usually consumes much processing time when a conventional computer is used. The main reasons are the vast amounts of pixel data that compose the image and the huge number of access times to the pixel data required for image processing. To cope with this situation, many unique architectures for image processing have been investigated and are utilized in various image-processing applications.

Considerable work has been done in Japan since 1970, especially for industrial applications, and considerable unique, basic, and applied research on industrial vision systems has been carried out.[8] The first successful application of image processing to industrial automation was the defect-inspection machine for printed-circuit boards[9] in 1972. In 1973, an automatic wire-bonding system with time-sharing vision was developed for transistor assembly.[10] This was later extended to development of systems for automatic assembly of ICs and LSIs.[11] A robot for bolting molds for concrete piles and poles, developed in 1974, was the first application to dynamic recognition of moving objects.[12] Through these pioneering studies, the importance of vision techniques is now recognized in many sectors. There were also some approaches done in the United States at that time. The first machine-vision trial was in the assembly of automobile tires.[13] However, image processing in this system was fully executed by a computer, which required an extremely long duration, preventing application of this system. The first successful application in the United States was in assembly of power ICs[14] in 1975.

Microcomputers did not exist at that time and the memory devices were expensive and small in capacity. Therefore, one important requirement was to process the image without using an image memory. The method of first storing the image in an image buffer memory, and then accessing the memory for processing, was unsatisfactory due to a poor cost/performance ratio. Therefore, the real-time image-processing method, in which processing is executed in synchronization with the image acquisition, was considered. In this method, the completion of image acquisition was meant to be the final stage of image processing. Thus, the technique of utilizing a two-dimensional local memory was studied extensively. Also, binary image processing became one of the main areas of study, and the combination of local memory and subsequent logic was extensively utilized. These formed the basis of local-parallel pipelined image processors, which are now on the market. In some applications, a minicomputer was utilized as a high-level controller of the image processor. However, the cost of the computer was still so high that a time-sharing group-controlled vision concept was applied[15] (see also Reference 10). In this concept, an image processor services a group of assembly machines, each equipped with a TV camera, and a number of image processors are hierarchically serviced by a single computer, thereby minimizing the computer cost per machine.

The appearance of microcomputers in the mid-1970s enabled higher-level control of vision systems. As their prices decreased, it became increasingly common to attach a microcomputer to every special-purpose image processor. Numerical and nonpixel-accessing functions in the image processor were then transferred to the computer. In the late 1970s, microprogram-controlled image processors were also realized for higher-speed applications. In and around 1980, higher-performance microprocessors, such as i8086 and M68000, became widely available, and higher-capacity memories such as 64K DRAM and 16K SRAM also became common. Thus, high-performance industrial vision having image memory was realized, and generalization of image processors was accelerated.

As the gate-array technique progressed and custom LSIs became more available in the mid-1980s, LSIs suitable for image processing began to appear.[16,17] A general-purpose image processor was realized, in which the preprocessing circuit was replaced with high-speed LSIs. The general-purpose image processor began finding wide application, as it could be adapted to many uses by changing the algorithms it could implement.

As described, the history of image processors for machine vision was greatly influenced by progress in semiconductor technology. In recent years, VLSIs for image processors have become available.[18] In the 1990s, new architectures for image processing ULSIs will be more intensively studied.

CHAPTER 2

Basic Algorithms for Machine Vision

2.1. Interimage Processing

2.1.1. Digitized Image

An image is, in general, a set of data that is spread two-dimensionally. The 2D space can usually be described by using a cartesian x-y coordinate system. An optical image such as a photograph can be regarded as a continuous image having data $f(x,y)$ for every pair of x and y coordinates. If the image is defined for only discrete x and y coordinates, these coordinates (x,y) are called picture elements, or simply pixels. Usually, the pixels are arranged so that each one corresponds to a crosspoint of a lattice, thus making the image a set of data arrayed two-dimensionally. A continuous optical image is usually converted to such a discrete image so that a computer or digital circuit can easily handle it. This transformation can be done by periodically sampling the image signal from an imaging device such as a TV camera with an appropriate sampling pitch. This can form an image sampled with a constant spatial pitch in both the x and y directions.

When the data $f(x,y)$ for the coordinate (x,y) can only be one of two logical values, "0" or "1," the image is called a binary image. When $f(x,y)$ can be an analog value within a certain brightness range, the image is called a gray-scale image. For ease of computer handling, the brightness range must be quantized to multibit data. Then the data $f(x,y)$ correspond to one of the digital levels, say, 256 levels from 0 through 255 for 8-bit quantization. This quantization of brightness is executed by an A/D converter circuit. The binary image is a result of extreme quantization into only two levels, sometimes called thresholding. The image in which the brightness is quantized and the coordinates are sampled is called a digitized image.

When a group of data is defined for each coordinate pair (x,y), the image is called a color image. In most cases, the group of brightness data for three principal colors, $r(x,y)$, $g(x,y)$, and $b(x,y)$, is defined for each pixel (x,y).

2.1.2. Algebraic Operations

When two digitized images are given, we can define the operations between two of their corresponding pixels. These operations can therefore be called interimage operations, and typically include addition and subtraction of images. Addition can be used to eliminate noises, where successively observed images of the same scene are totalled and averaged. Subtraction can be used to find portions of the image that are different from each other. For example, by subtracting a complicated background image from the input image, including an object, an image portion of the object can be extracted. From the difference between the two images, observed at times $t - dt$ and t, it is possible to obtain an image portion of a moving object that moves within the time interval dt. Thus, interimage operations are, in general, image-to-image transformations that generate one output image from two or more images.

2.1.3. Logical Operations

When two given images, $f_1(x,y)$ and $f_2(x,y)$, are both binary, we can define logical operations, such as AND and OR between them. In particular, when one image $f_2(x,y)$ is a known window pattern, the logical multiplication (AND) extracts a portion of the image f_1, as shown in Figure 9. This process is called masking, that is, f_2 masks f_1. In this case, f_1 may also be a gray-scale image, so the masking process can be executed by gating the gray-scale f_1 image with a gate-switching signal f_2.

The logical summation (OR) is used to generate a mixed image of f_1 and f_2, where the pixel value is "1" when either $f_1(x,y)$ or $f_2(x,y)$ is "1." The negation operation (NOT) generates a negative binary image whose pixel values are the reverse of the original image. These logical operations are basic operations in binary im-

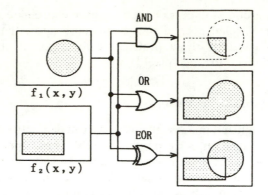

Figure 9. Basic logical operations.

age processing and are widely utilized in every aspect of image processing.

The exclusive-OR operation (EOR) outputs an image with a logical pixel value of "1," where $f_1(x,y)$ and $f_2(x,y)$ are different. Therefore, it is possible to evaluate to what degree the two images coincide. Considering one image f_2 as a standard pattern, and executing EOR with the other image f_1 by shifting their relative positions, we can search within the image f_1, the portion with the best match to the standard pattern f_2. In this case, the outside portion of the finite-sized standard pattern is regarded as an undefined "don't-care" portion and is excluded from the EOR operation. This method is extensively utilized as a pattern-matching method that aims at recognition of types and positions of objects.

2.2. Filtering

2.2.1. Linear Filtering

Filtering is a typical intraimage operation in image processing for converting an image $f(x,y)$ into a different image $g(x,y)$. In this operation, the output pixel value $g(x,y)$ is obtained not only by the corresponding pixel value $f(x,y)$ of the input image, but also by other neighboring pixels within the region A, as shown in Figure 10. Thus, the operation is called spatial filtering in contrast to temporal filtering in signal processing.

Now we assume that images $f(x,y)$ and $g(x,y)$ are gray-scaled. For ease of processing, region A of the filtering operation is generally selected as a rectangular region. That is, when a coefficient matrix a_{ij}, as shown in Figure 11, is considered as an $m \times n$ rectangular pixel region, the output image at the pixel (x,y) is given by

$$g(x,y) = \sum_{j=1}^{n} \sum_{i=1}^{m} a_{ij}f(x+i-1,y+j-1)$$

(2-1)

This is a "multiplication-and-addition" operation between the coefficient matrix a_{ij} and the image $f(x,y)$ and is executed as illustrated

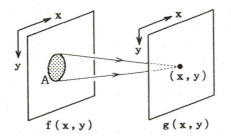

Figure 10. Filtering operation.

a_{11}	a_{21}	a_{31}		a_{m1}
a_{12}	a_{22}	a_{32}		a_{m2}
a_{13}	a_{23}	a_{33}		a_{m3}
a_{1n}	a_{2n}	a_{3n}		a_{mn}

Figure 11. Coefficient matrix.

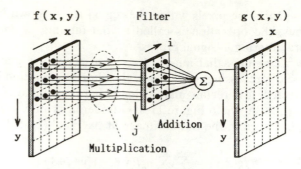

Figure 12. Digital filtering.

in Figure 12. The coefficient matrix is called a spatial operator, or simply a filter. Because the filtering described by Eq. (2-1) is a linear combination of the pixels in region A, it is called linear filtering. Pixels $f(x,y)$ and $g(x,y)$ correspond when $i = 1$ and $j = 1$. Therefore, the leftmost and uppermost pixel of the $m \times n$ filter is regarded as the basepoint of the filter. To make the center of the filter the basepoint, Eq. (2-1) can be replaced by

$$g(x,y) = \sum_{j=1}^{n} \sum_{i=1}^{m} a_{ij} f(x+i- \frac{m+1}{2}, y+j- \frac{n+1}{2})$$

$$(2\text{-}2)$$

where m and n are odd. The only difference between Eqs. (2-1) and (2-2) is the write-in position of the resulting pixel image in Figure 12. Therefore, there is no intrinsic difference between these two filterings except that the output images are shifted from each other.

Examples of linear filters are shown in Figure 13, where Figure 13(a) represents a filter that adds all neighboring pixels with equal weighting. Thus, it averages the neighboring pixels, producing a less noisy smoothed image. Filters in Figures 13(b) and 13(c) are for spatial differentiation of the input image. They extract the edge portion, where the brightness gradient in the respective x and y directions is large. To obtain a gradient image by using first-order spatial differentiation, two images, one filtered for x and the other filtered for y, must be combined. To simultaneously extract edge

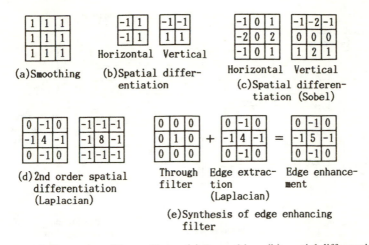

Figure 13. Examples of linear filters. (a) Smoothing, (b) spatial differentiation, (c) spatial differentiation (Sobel), (d) second-order spatial differentiation (Laplacian), and (e) synthesis of edge-enhancing filter.

information in the x and y directions, a second-order spatial differentiation, that is, a Laplacian filter, can be applied. Examples of Laplacian filters are shown in Figure 13(d).

To execute the filtering operation by computer, the process of accessing a number of pixels, multiplying them by filtering coefficients, and adding the results of the multiplications should be repeated for each of the pixels in the image. This takes too much time. To speed up the process, a local-parallel processing hardware can be considered, as shown in Figure 14. In this architecture, $m \times n$ regional pixels are extracted in parallel by furnishing $n - 1$ shift registers, each having delay elements of M pixels, the same as the number of horizontal pixels in the image, as well as by furnishing $(m-1) \times n$ one-pixel delay elements. This can be called a two-dimensional local memory. It outputs $m \times n$ pixels in parallel each time one pixel is input and shifted. The outputs can be used for multiplication and summation calculations in subsequent circuits. Thus, filtering can be achieved in real time in sychronization with the input image signal.

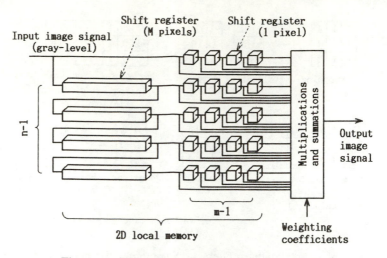

Figure 14. 2D local-parallel digital filtering circuit.

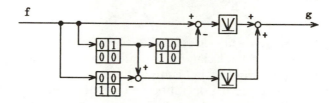

Figure 15. Block-diagram representation of filtering (Forsen filter).

2.2.2. Nonlinear Filtering

Many nonlinear filters are available for image processing as well as the linear filter just described. These filters include nonlinearity in pixel processing. One typical 2×2 filter converts the image $f(x,y)$ into $g(x,y)$ as described by the equation:

$$g(x,y) = |f(x,y) - f(x+1,y+1)|$$
$$+ |f(x,y+1) - f(x+1,y)| \qquad (2\text{-}3)$$

This is effective for deriving edge images and is called a Forsen filter. In general, filtering flow can be represented by a block diagram. The filter of Eq. (2-3) is represented by the block diagram shown in Figure 15.

There are other nonlinear filters, including a median filter, which outputs a value at an intermediate pixel when the pixels in the $m \times n$ filtering window are arranged in a line in size order. It is a type of edge-preserving smoothing filter that makes the image smooth yet avoids blurring at the edges.

2.2.3. Recursive Filtering

A nonlinear filter that utilizes its output recursively by feeding it back into its input is called a recursive filter. However, conventional filters, as described in the previous sections, do not feed back and are called nonrecursive filters. The simplest and most useful recursive filter is the propagation filter. One such example is represented by the equation:

$$g(x,y) = f(x,y) \vee \{g(x-1,y) \wedge g(x,y-1)\} \qquad (2\text{-}4)$$

where $\vee$ denotes logical summation (OR), and $\wedge$ denotes logical multiplication (AND). The scanning mode of the input image of this filter is the same as conventional TV scanning: horizontal rightward first and then downward. By this filtering operation, the logical "1" portion in the original image $f(x,y)$ propagates rightward down, making the rightmost bottom corner of the portion straight, as shown in Figure 16.

There is another type of recursive filter, which can be called a distance-transform filter. This filter usually outputs a larger value at the inner pixel within a connected figure, depending on the distance of the pixel from the figure edge. The pixels to which the

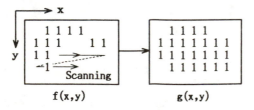

Figure 16. Propagation filter.

large values are given are far from the edge, thus facilitating extraction of medial lines from the figure.

As an example of the distance-transform filter, consider the following algorithm:

$$g_1(x,y) = [\max \{g_1(x-1,y),g_1(x-1,y-1),g_1(x,y-1)\} + 1] \wedge f(x,y)$$
$$g_2(x,y) = [\max \{g_2(x+1,y),g_2(x+1,y-1),g_2(x,y-1)\} + 1] \wedge f(x,y)$$
$$(2\text{-}5)$$

where $\wedge$ denotes the logical multiplication (AND) between a binary value and each bit of a multilevel value. The algorithm generates two images, g_1 and g_2, as shown in Figure 17. Here larger values are given to pixels from upper left to lower right in g_1, and from upper right to lower left in g_2, depending on their lattice distances. In this case, the scanning in the x direction is reversed for g_2 generation. Thus, the two images can be combined as follows:

$$g(x,y) = \max \{g_1(x,y), g_2(x,y)\} \qquad (2\text{-}6)$$

This is the type of interimage operation described earlier. It generates an image having larger values (lattice distances) at the bottom portion of a figure. Thus, it can serve as a size filter, which can extract only larger figures (blobs). This can be executed by thresholding the image $g(x,y)$ to extract nucleus portions of larger figures, which can then be dilated within the original image $f(x,y)$. This constrained broadening operation is very important for extracting

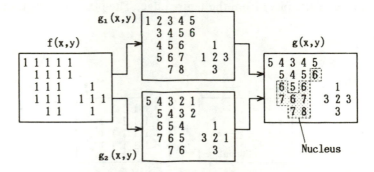

Figure 17. Size filter based on distance transform.

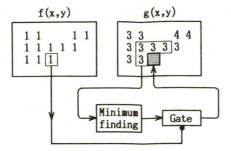

Figure 18. Labeling filter.

real defects among many candidates in the visual-inspection machine[19] and will be described later in more detail.

Another well-utilized recursive filter is a labeling filter for binary images. This filter puts the same number to the pixels forming the same figure, thus producing an image in which each figure has different numbering. This is useful as a preprocessing scheme to discriminate each object in the image for separation. As illustrated in Figure 18, the filter designates a number incrementally each time an unconnected pixel appears in the scanning process. When the pixel is connected to its four neighboring pixels, the minimum number of these four is chosen for that pixel. If a figure is upward-concave, it consists of more than one type of pixel having different numbers. Therefore, if the neighboring pixels possess different numbers (3 and 4, for example), the numbers are memorized in a list representing that these numbers are the same. By using this list, the output image can be renumbered at a later scanning, depending on requirements.

2.3. Morphological Operation

Magnification, reduction, and rotation of a whole image can usually be executed by the well-known Affine transform. Dilation and erosion of an object figure contained in the image can be processed by logical operations when the image $f(x,y)$ is binary. For example, dilation can be achieved by the following logical summation of the plural number of the input pixels around (x,y):

$$g(x,y)= \bigvee_{i,j}f(x+i,y+j) \qquad \text{for}$$
$$-(m-1)/2 \le i \le (m-1)/2$$
$$-(n-1)/2 \le j \le (n-1)/2 \qquad (2\text{-}7)$$

where m and n are odd numbers and correspond to the size of dilation in the x and y directions. The following logical multiplication operation gives the eroded image $g(x,y)$:

$$g(x,y)= \bigwedge_{i,j}f(x+i,y+j) \qquad (2\text{-}8)$$

These dilation and erosion operations, using logical filtering, can be called morphological operations.

To execute such operations dynamically in real time by synchronizing with the input image signal, the local-parallel processing hardware already described can also be utilized. The hardware for binary image processing is represented by the circuit of Figure 19. It consists of a two-dimensional local memory and a logical circuit. The two-dimensional local memory consists of $n-1$ shift registers, each having M delay elements and $(m-1) \times n$ one-pixel delay elements, and derives a rectangular $m \times n$ pixel area in parallel. The logical circuit applied to the $m \times n$ parallel data can achieve a characteristic operation depending on the types of logic. For example, the dilation described before can be achieved by adding a logical OR circuit, as shown in Figure 19. In this case, the output

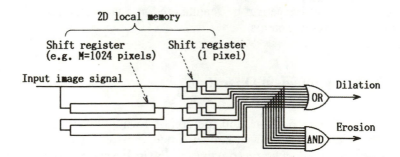

Figure 19. 2D local-parallel logical filtering circuit (dilation and erosion).

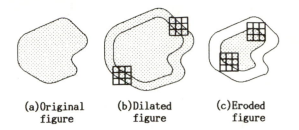

(a)Original figure (b)Dilated figure (c)Eroded figure

Figure 20. Morphological operation of object figures. (a) Original figure, (b) dilated figure, and (c) eroded figure.

image can be logical "1" when the window, having a size of $m \times n$, partially overlaps with an object figure in the image, as shown in Figure 20. The erosion can be executed by an AND circuit. In this case, the $m \times n$ window must be fully overlapped with the object figure to output the logical "1." When the output of an imaging device, such as a TV camera, is connected to the two-dimensional local memory circuit, and when the outputs of the logical circuits are connected to a display device, the dilated and/or eroded images generated in real time can be observed.

2.4. Extraction of Patterns

2.4.1. Small Pattern Extraction

The dilation and erosion operations described before can be utilized to extract small portions from a binary pattern, as shown in Figure 21. A small portion in the pattern is filled when we apply a dilation operation first and the erosion operation next. Then the erosion and dilation operations are again applied to remove the small portion in the background portion. The output image signal is then compared with the original input signal by utilizing an interimage operation, in this case, exclusive-OR. Thus, only small portions can be extracted. This method is especially effective for finding defects in inspection applications (see Reference 9).

Extraction of true particles from a gray-level image containing

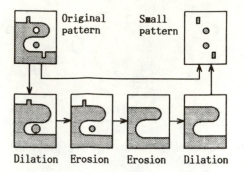

Figure 21. Small portion extraction by morphological operations.

many particlelike artifact patterns is often important in industrial vision systems. It is sometimes very difficult, especially when brightness differences between the particle and the artifacts in the image are subtle. One method utilizes two thresholding values. The large threshold provides a nucleus image of particles, $f_1(x,y)$, as these particles usually contain a highlighted portion. Thresholding at a lower value gives a control image, $f_2(x,y)$, containing all particles and some artifacts. Thus, the figures in image $f_1(x,y)$ are dilated within the inner portion of corresponding figures in $f_2(x,y)$, resulting in particle images whose original sizes have been restored. This operation can be called restricted broadening (see Reference 19). It is useful for particle analysis and defect inspection. It can also be executed in a real-time mode by using the circuit shown in Figure 22. The shift register for f_2 is used to synchronize its pixel position with that of f_1.

To extract a small portion from a gray-level image, maximum-value and minimum-value filters can be applied. These are nonlinear filters that output maximum and minimum values among the pixels in a filtering window. They are expressed by

$$g(x,y) = \max_{i,j} f(x+i,y+j)$$
$$g(x,y) = \min_{i,j} f(x+i,y+j) \qquad (2\text{-}9)$$

where i and j are as defined in Eq. (2-7). The effects of these filtering operations are illustrated in Figure 23 for the one-

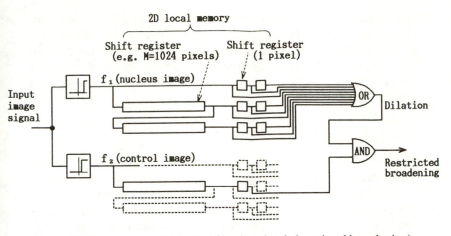

Figure 22. 2D local-parallel logical filtering circuit (restricted broadening).

Figure 23. Maximum- and minimum-value filtering. (a) Original image, (b) maximum-value filtering, (c) minimum-value filtering, and (d) subtracted image.

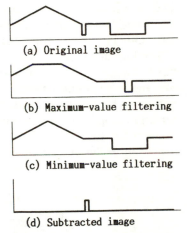

(a) Original image

(b) Maximum-value filtering

(c) Minimum-value filtering

(d) Subtracted image

dimensional case. The maximum-value filter fills in small brightness dips, and the minimum-value filter eliminates small portions with upward brightness protrusion. These two filters correspond to OR and AND filters for binary images. Maximum-value filtering is first executed, as shown in Figure 23(b), and then minimum-value filtering is applied to the result, as shown in the Figure 23(c),

eliminating local variations. Thus, the final output image only contains the components of macroscopic brightness variation contained in the original image. Subtraction of this output image from
the original image yields an image containing only objects of less
than certain size. Thus, small-object images can be extracted, even
though the input image is shaded.

2.4.2. Specific Pattern Extraction

To extract a specific pattern from a binary image, an appropriate
logic circuit can be used to process the pixel output from the two-
dimensional local memory. To extract a corner pattern, for example, the circuit shown in Figure 24 can be used. This circuit outputs
logical "1" only when the leftmost and uppermost pixels are all "0"
and the lower-right pixels are all "1." The intermediate area is
called the "don't-care" portion, and is neglected in the decision.
The preparation of this "don't-care" portion is often effective in
absorbing the digitizing errors that are liable to occur at the pattern boundary due to the image-sampling effect.

The logical circuit just described is rather hard-wired, and is not
flexible when more patterns are to be extracted. To make it flexible, the logic circuit shown in Figure 25 can be utilized, where a
pattern to be detected is stored in a two-dimensional pattern register and compared with two-dimensional local patterns extracted
sequentially by the two-dimensional local memory. The comparison is an exclusive-OR operation between corresponding pixels,

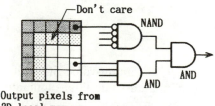

Output pixels from
2D local memory

Figure 24. Extraction of corner pattern.

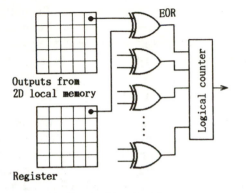

Figure 25. Extraction of arbitrary patttern.

and the results are counted logically to yield a value representing the pattern difference. Therefore, the pattern is regarded as coincident when the value becomes zero in an ideal case or minimum in an actual case. This process, in which patterns are directly compared, is called pattern matching, and the pattern to be detected, which is stored in the register in this case, is called a standard pattern or template.

2.5. Extraction of Features

2.5.1. Frequency Distribution

To extract macroscopic features of an image, a frequency distribution is often derived. For example, a histogram for a whole image representing the number of pixels versus brightness gives the optimum thresholding value for binarization, especially when it has twin peaks: one for the object area and the other for the background.

The frequency distribution can be effectively combined with the direction-coded image. The direction-coded image can be obtained by the following two steps. First, a gradient-vector image is calculated by the first-order differentiation of the original gray-level image in both x and y directions. Then the unit vector whose direction is perpendicular to the gradient vector is calculated at each pixel. The direction-coded image is a set of such unit vectors. The following indicates the steps:

$$\mathbf{V}(x,y) = [\partial f(x,y)/\partial x, \; \partial f(x,y)/\partial y]$$
$$g(x,y) = \text{Arg}\,\{\mathbf{V}(x,y)\} + \pi/2 \qquad (2\text{-}10)$$

where the Arg operator denotes the calculation of angular deviation of the gradient vector $\mathbf{V}(x,y)$. The output image $g(x,y)$ is called a direction-coded image. An example is schematically shown in Figure 26. This image is invariant to the change in contrast a and to the change in mean brightness level b in the original image variations, $af(x,y) + b$, caused by illumination change.[20] Furthermore, each pixel has a minimal geometric meaning: the tangential direction of the brightness boundary, in contrast to the original brightness image. Therefore, a histogram of pixels describes the dominant direction, enabling detection of linear contours. For a rectangular object, the histogram can be folded with a $\pi/2$ period to give it a more dominant peak than the original histogram, as shown in Figure 27, enabling detection of orientation θ_0 of the rectangular object. This method can be extended to the image of piled boxlike objects observed obliquely from above[21] (see also Reference 20). Other line-detection methods include a Hough transform and line fitting with least-square error.

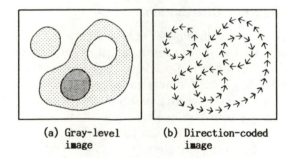

(a) Gray-level image

(b) Direction-coded image

Figure 26. Transformation into direction-coded image. (a) Gray-level image and (b) direction-coded image.

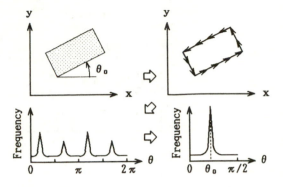

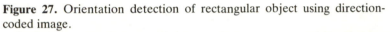

Figure 27. Orientation detection of rectangular object using direction-coded image.

2.5.2. Projection Distribution

Projection distribution can be defined as a curve along a baseline to which brightness of pixels is projected perpendicular to the baseline and summed. When the image is binary, the distribution gives width information of the object in the projected direction. The following equations represent the process when the baselines are x and y axes:

$$g_1(x) = \sum_y f(x,y)$$
$$g_2(y) = \sum_x f(x,y)$$

(2-11)

Projection distribution is frequently utilized when an object position in the image is identified. When only one object is allowed in an image, these distributions in both x and y directions are utilized to find the extreme position of the distributions and the averaged center position (x_0, y_0), as shown in Figure 28. When objects are sure to be arranged horizontally, separation of area in the x direction and detection of average x positions for each subarea can be first achieved by the distribution projected to the x axis. Then, y positions are detected by the distribution projected to the y axis for each separated subarea. When such separation is difficult, the labeling process described earlier can be applied.

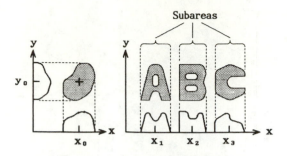

Figure 28. Projection distribution.

The projection-distribution curves often contain very important information on the object shape. Therefore, they can be transformed into Fourier series, and their coefficients can be regarded as feature parameters for a subsequent shape-recognition process.

2.5.3. Feature Parameters

To extract contour information from a solid figure in a binary image, the method shown in Figure 29 can be applied.[22,23] In this method, 3×3 pixels extracted in real time from the two-dimensional local memory are checked to see if $f(x-1,y) \neq f(x+1,y)$ or $f(x,y-1) \neq f(x,y+1)$ when $f(x,y) = 1$. If this condition is satisfied, pixel (x,y) is on the outermost boundary of the object figure, as shown by the circles in Figure 30. In this figure, the double-circle symbols specifically indicate boundary pixels in both x and y directions. These two types of pixels can be counted in real time, and the resulting values, B and C, are used to calculate the approximate length of contour line L as follows:

$$L = (B^2 + C^2)^{1/2} \tag{2-12}$$

This will be obvious from the figure. Area A of the object can also be obtained easily by counting the logical "1" pixels. By using two feature parameters, A and L, a compound-feature parameter, L^2/A, can be defined. This parameter becomes a constant, 4π,

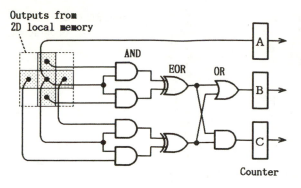

Figure 29. Extraction of feature parameters.

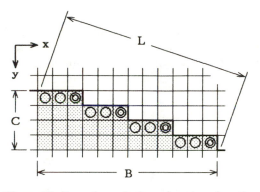

Figure 30. Boundary pixels and contour length.

when the object is circular. Thus, it can be utilized as a simple yet very powerful parameter for detecting the degree of deformation of intrinsically circular objects, thereby distinguishing between normal objects and defective objects having cracks.

More general feature parameters can be derived by using chain-coded data. A solid figure can be represented by closed chain-coded data, as shown in Figure 31. Transformation of a binary image into chain-coded data can be achieved in a one-pass mode by a sequential machine in real time, utilizing only two scanline

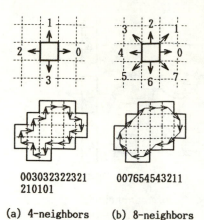

Figure 31. Chain code: (a) 4-neighbors and (b) 8-neighbors.

003032322321
210101

007654543211

(a) 4-neighbors   (b) 8-neighbors

data at one time.[24] This eliminates a conventional, lengthy, edge-following operation. The calculation of many feature parameters, such as area, peripheral length, and moment of inertia, can be easily executed by directly using chain-coded data.

CHAPTER 3

Recognition of Shapes and Positions

3.1. Pattern-Matching Method

3.1.1. Distance between Patterns

The distance $D(i,j)$ between two digitized images, $f_1(x,y)$ and $f_2(x,y)$, both of which are gray-level images, is defined as

$$D(i,j) = \sum_y \sum_x [f_1(x+i,y+j) - f_2(x,y)]^2 \tag{3-1}$$

That is, the two images are shifted relative to each other by (i,j) in the two-dimensional image plane. The differences are taken between corresponding pixels, squared, and then summed. It may be easier to understand the word "distance" if the right-hand side of the equation were square-rooted. However, here we define "distance" as it appears in the equation.

The coordinate (i,j), where distance $D(i,j)$ is minimum, is the relative position where two images best match each other. When the two images are exactly the same, i.e., $f_1(x,y) = f_2(x,y)$ for all (x,y), Eq. (3-1) gives $D = 0$ at $i = j = 0$. If f_1 is an input image and f_2 is a standard pattern, the process of finding position (i,j) of f_1 relative to the standard pattern f_2, where $D(i,j)$ is a minimum, is called pattern matching. The standard pattern f_2 is usually defined as one with a finite size of $m \times n$ pixels. f_2 is not defined outside this area, and the calculation of Eq. (3-1) is excluded for these areas. Thus, the process is equivalent to that of finding where the standard pattern exists within the input-image plane.

3.1.2. Correlation of Patterns

Distance $D(i,j)$ in Eq. (3-1) can be expressed as follows:

$$D(i,j) = \sum_y \sum_x f_1^2(x+i,y+j)$$

$$-2 \sum_y \sum_x f_1(x+i,y+j)f_2(x,y)$$

$$+ \sum_y \sum_x f_2^2(x,y)$$

$$(3\text{-}2)$$

The first and third terms can be neglected from the minimum-value finding when these are considered constant. From the second term, we get

$$C(i,j) = \sum_y \sum_x f_1(x+i,y+j)f_2(x,y)$$

$$(3\text{-}3)$$

As the sign is reversed, the operation of finding the minimum of $D(i,j)$ becomes equivalent to finding the maximum of $C(i,j)$. $C(i,j)$ is called a correlation, and the position (i,j) that gives the maximum correlation $C(i,j)$ is the best matched position. When $f_2(x,y)$ is defined only in the area of $m \times n$, this equation represents the same operation as the linear spatial filtering described earlier. Thus, filtering is the process of calculating the correlation between the coefficient matrix of the filter and the image. In actual applications for finding the best matched position, the correlation in Eq. (3-3) must be normalized by using variances and averages of f_1 and f_2.

3.1.3. Binary Pattern Matching

When images $f_1(x,y)$ and $f_2(x,y)$ are both binary, the distance between the two images, $D(i,j)$, can be represented by replacing the subtraction operation with the exclusive-OR operation as follows:

$$D(i,j) = \sum_y \sum_x f_1(x+i,y+j) \oplus f_2(x,y)$$

$$(3\text{-}4)$$

where $\oplus$ denotes the exclusive-OR, and Σ denotes the algebraic summation of logical "1" pixels, thus yielding the total number of differing pixels between the two images. The position (i,j) where the $D(i,j)$ is a minimum is the best matched position. In general, f_2 is a standard pattern defined for a certain $m \times n$ area, and the pattern-matching process finds the best matched position from f_1. In this case, distance D is equivalent to the degree of mismatching and is used as a certainty index. It is also noted that the algebraic summation of the negation of the right-hand term of Eq. (3-4) gives the degree of coincidence.

A typical hardware furnished with this binary pattern-matching method (see Reference 10) is shown in Figure 32. The input image from a TV camera is thresholded into a binary image and fed into a two-dimensional local memory. From this local memory, the two-dimensional local patterns are sequentially output, synchronizing with the scanning of the TV camera. A pattern-matching circuit consisting of exclusive-ORs and a logical counter is located between the output pixels from the local memory and the output pixels from the register storing a standard pattern. The circuit outputs information on the degree of coincidence. When the degree of coincidence is larger than that previously memorized, the register is updated. At the same time, a gate is opened and the positional data is stored into a coordinate register. Thus, after a

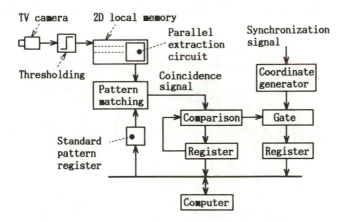

Figure 32. Configuration of pattern-matching processor. (From Reference 8.)

vertical scan of the TV camera, the position, as well as the degree of coincidence where the standard pattern is best matched, is settled in each register. These data are then received by a microcomputer within the blanking period of the TV camera. Within the same period, the standard pattern in the register can be replaced by another standard pattern, making it possible to find the position of different patterns in successive TV frames.

Thus, when we select a plural number of unique patterns from the object image, as shown in Figure 33, and store them in the computer as standard patterns, the positions of the patterns can be detected sequentially, one for each vertical scan. The absolute position of an object in a two-dimensional plane can be determined from the position of only one pattern when the object does not have any rotational error. However, to check whether the pattern is correctly detected, one more pattern position must be detected. As the relative position and orientation between two standard patterns are known, a geometric test of distance d_{12} and angle θ_{12} between the two detected positions can be used to judge whether both positions are correctly detected and can be achieved within the blanking period of the TV scan. This method is being utilized as a visual aid for wire-bonding machines in the assembly of transistors, ICs, and LSIs to automatically detect the position of each chip fed into the machines. An example of a wire-bonding machine is

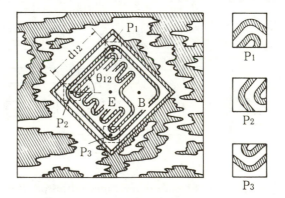

Figure 33. Object image and local patterns.

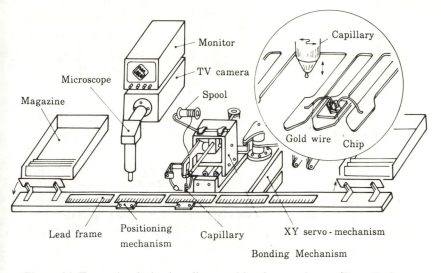

Figure 34. Example of wire-bonding machine for transistors. (From Reference 8.)

shown in Figure 34. In most cases, the first two patterns are enough to find the object position, thus consuming two vertical scanning periods (33 ms). However, when two patterns are not enough to satisfy the geometric test condition, a third pattern must be checked. The test is then executed between the first and third patterns and then between the second and third patterns. If the test is still not satisfactory, subsequent standard patterns are searched repetitively until the object position is fixed. Meeting the test condition implies detection of the correct pattern pair, and the target-electrode position can be calculated by using known data of relative positions among the standard pattern positions and the electrode positions. In general, the position and orientation are variable for various fed-in conditions. Therefore, standard patterns can also be selected from object images with a few orientation errors.

Now, suppose that we prepare m standard patterns. For all patterns, we assume a constant recognition rate p. The probability q_m that object position detection first succeeds at the mth matching is represented by

$$q_m = (m - 1)p^2(1 - p)^{m-2} \tag{3-5}$$

The probability r_m that position detection is terminated by the mth matching can be represented as the accumulated probability of q_m for $m = 1$ (actually, $m = 2$ as $q_1 = 0$) through $m = m$. Thus:

$$r_m = 1 - (1 - p)^{m-1}(1 - p + mp) \tag{3-6}$$

Even if the recognition rate p for detecting a single pattern is rather low, the resulting position-detection probability r_m increases rapidly as m increases. For example, for $p = 0.9$ and $m = 2$, both patterns must be correctly detected, thus yielding lower probability $r_2 = 0.81$ ($= 0.9 \times 0.9$). However, when $m = 3$, two patterns out of three can be correctly found, thus giving a higher probability of $r_3 = 0.972$. When $m = 9$, we obtain $r_9 = 0.9999999$. This theoretically high probability is one feature of this method, which can be called a multiple local pattern-matching method.

3.1.4. Automatic Selection of Standard Patterns

Each standard pattern selected from the object pattern should be unique to minimize incorrect matching. Incorrect matching can occur if a pattern similar to the standard pattern occurs at a different position. Suppose we select a local pattern at (x_0,y_0) and compare this pattern to another pattern at an arbitrary point (x,y). The total number of differing pixels between these two patterns can be defined as a matching error, $E(x,y)$. That is,

$$E(x,y) = \sum_j \sum_i f(x+i,y+j) \oplus f(x_0+i,y_0+j)$$
$$0 \le i \le m - 1 \quad \text{and} \quad 0 \le j \le n - 1 \tag{3-7}$$

where m and n are the horizontal and vertical size of patterns, respectively; $\oplus$ denotes the exclusive-OR; and Σ denotes the algebraic summation of logical "1." The evaluation index E_c for the selection of the standard pattern becomes:

$$E_c = \min_{(x,y)\in S} E(x,y) \tag{3-8}$$

where S is the area excluding (x_0,y_0) and its vicinity. When the evaluation index becomes small, the potential for incorrect matching becomes high. Therefore, a pattern with a high index must be chosen for the standard pattern. An optimum pattern position can be searched by evaluating E_c for every (x_0, y_0). However, a large number of operations is required for such a search, so the search area is limited. Positions (x_0, y_0) at which local patterns satisfy certain criteria are flagged and the search is executed only for these positions. These criteria are

1. The ratio of black to white pixels is nearly equal to one (excluding portions containing little information).
2. The pattern contains simpler black-and-white subpatterns (excluding portions containing mainly lines, as line patterns easily lose their original shape due to the variations in thresholding conditions).
3. The length of the black-and-white boundary is within an acceptable range (excluding portions where the pattern is too simple or too complicated).
4. The pattern contains both perpendicular boundaries (excluding portions with only one-directional patterns).

To apply these criteria, a pattern is first decomposed into 2×2 patterns. The number of 2×2 patterns within the original $m \times n$ pattern is $(m - 1) \times (n - 1)$. They are then grouped as shown in Figure 35, and their frequencies, classified into one of the groups G_1, G_2, G_3, or G_4, are counted. Normalized frequencies are denoted as d_1, $d_2(= d_{21} + d_{22})$, $d_3(= d_{31} + d_{32})$, and d_4, where $d_1 + d_2 +$

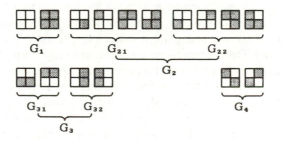

Figure 35. Grouping of 2×2 pixel patterns.

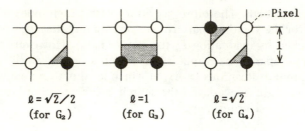

Figure 36. Edge lengths of 2 × 2 pixel patterns.

$d_3 + d_4 = 1$. A vector **D**, where **D** $= (d_1, d_2, d_3, d_4)$, can be regarded as a feature vector representing one aspect of the pattern characteristics. Criterion 2 stated before can be judged by checking if d_1 is within a certain range. Also, criterion 4 can be judged by checking if both d_{21} and d_{22} are larger than a certain value, or if both d_{31} and d_{32} are larger than a certain value. Since the pattern edge length of each group of 2 × 2 patterns can be represented as shown in Figure 36, the total length of boundaries within the pattern can be approximated by $\sqrt{2}d_2/2 + d_3 + \sqrt{2}d_4$. This scalar value represents the complexity of the pattern, and criterion 3 can be judged by checking if this value is within a certain range. By using these criteria, the area to be evaluated is easily reduced to less than 20% of the original image area. As the search using Eqs. (3-7) and (3-8) usually requires a number of operations proportional to the image size to the fourth power, this reduction is extremely effective.[25]

When a plural number of standard patterns is automatically selected in sequence, the pattern with maximum E_c is first registered. To avoid the situation where a number of patterns is selected from a small concentrated area, the distances to preregistered patterns are evaluated. Only when these distances are large enough, are the patterns registered as standard patterns. This enables the automatic selection of a plural number of standard patterns spread over the object image. An example of such a selection is shown in Figure 37 for an LSI chip image. It is easily achieved by using matching hardware with a two-dimensional local memory, described earlier.

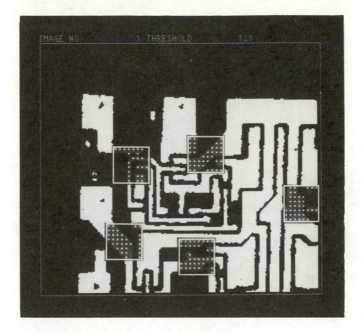

Figure 37. Automatic selection of standard patterns. (From Reference 25.)

3.2. Pixel-Counting Method

3.2.1. Shape Classification

The shape of objects for classification is detected based on black-and-white areas within windows generated in the image plane. For example, silhouetted images of different bottles can be observed through two windows set around bottle-shoulder level, where shape differences are most notable. This is illustrated in Figure 38. Areas a and b are counted for each window fixed to the image plane, when the bottle is positioned at the center of the image plane. It is also possible to dynamically detect a bottle moving from left to right in the image plane. The instant area a coincides with area b, the bottle is at the center position, and the area $a + b$ can be used for shape classification.

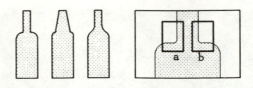

Figure 38. Shape recognition by pixel-counting method.

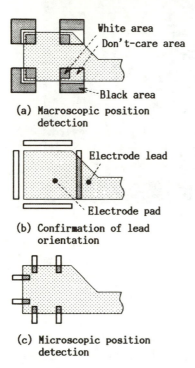

(a) Macroscopic position
    detection

(b) Confirmation of lead
    orientation

(c) Microscopic position
    detection

Figure 39. Shape and position recognition of LSI electrode. (a) Macroscopic position detection, (b) confirmation of lead orientation, and (c) microscopic position detection.

This method was first applied to recognition of types of electrodes in LSI assembly (see Reference 11). As shown in Figure 39, four corners of an electrode are first found, based on the pattern-matching technique described earlier, where four types of standard patterns with "don't-care" portions are used. Four windows are generated around the electrode by using its detected rough position. The window through which most of the electrode is seen gives

the lead direction information, confirming that the desired electrode is correctly detected.

3.2.2. Position Detection

The pixel-counting method can also be applied effectively to the
detection of positions. In the previous example, smaller windows
are generated in response to the shape and rough position of the
electrode. A more precise position is then determined from the
area the electrode occupies in these windows. This is an example of
a hierarchical position-recognition method combining macroscopic
and microscopic techniques.

The principle of position recognition based on pixel counting is
rather simple. Windows are put on either side of an object and
differential output is taken from them. To detect a circular hole
position, for example, four windows are generated so that two of
them are at opposite sides of the hole in the x direction and the
other two in the y direction, as shown in Figure 40. From the
output pixel numbers, a, b, c, and d, for each window, position
(x,y) of the hole center is determined in relation to the image
center (x_0,y_0) by

$$x = x_0 + (a - b)/2w \qquad y = y_0 + (c - d)/2w \qquad (3-9)$$

where w is the width of the windows. This method was utilized in
robot vision for automatic water-pressure testing of pumps, as
illustrated in Figure 41. The positions of the inlet and outlet holes
of the pumps are recognized, and hoses are automatically connected to them.[26] Quarter segments of a circle are first used for

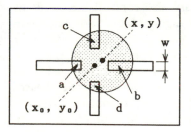

Figure 40. Principle of position recognition by pixel-counting method.

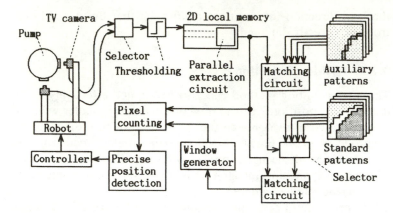

Figure 41. Vision-based hose-connecting robot for pressure testing of pumps.

pattern matching to detect a rough position to capture the hole, even though it may be partly out of the field of vision. Once the hole position is captured, four windows are generated and the previous position calculation is executed for every scanning field. Thus, the method used here is also hierarchical. The TV camera is mounted at the tip of the robot arm and the position information obtained by each vertical scan is used to control the arm to bring the hole center into the center of the camera's view field. This technique is called visual feedback.

The window positions can also be controlled by this method. This facilitates tracking of a moving object within a fixed field of vision. If the camera itself is also movable, a large displacement of an object can be followed with the object still in the view field. Sizes and shapes of the windows are also controllable, enabling tracking of a moving and shape-changing object and continuous outputting of its average position.

The pixel-counting method can also be applied for inspection as a simple detector of abnormality. Examples include recognition of the existence of labels and their positional errors on bottle surfaces in a labeling process used in medicine and chemical industries. Problems in application of this method are to obtain a good contrasting image between the object and its background and to stably threshold it into a binary image. These are the keys to successful application.

3.2.3. Recognition of Moving Objects

A machine-vision system based on the pixel-counting method can detect an object that traverses the image field from left to right. The vision system outputs the type of object and the timing of its passage. This was used for vision of a bolting robot (see Reference 12) that finds and fastens (or loosens) bolts arranged side by side on the flange of a mold for concrete piles and poles. The robot is shown in Figure 42. It was the first application of machine vision to dynamic recognition of moving objects.

A number of windows, each having four subareas, are set at fixed positions in the image, corresponding to the characteristic portions of the object to be detected, as shown in Figure 43(a). The combined width of two central subareas is chosen to be equal to the object width. When the object comes into the center of the window, the two subareas are filled by the object and the other

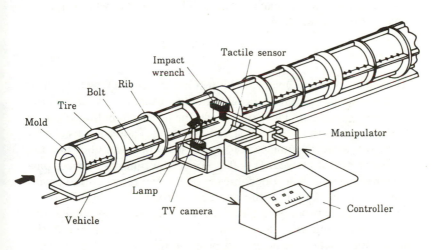

Figure 42. Bolting robot.

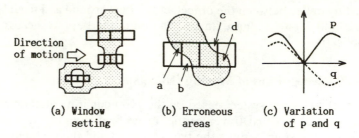

Figure 43. Recognition of moving object. (a) Window setting, (b) erroneous areas, and (c) variation of p and q.

two outer subareas are covered by background. At this instant, therefore, two signals should be output, one signaling that the width of the unknown object coincides with the predetermined width, and the other signaling that now is the time of passage.

To do this, erroneous areas a, b, c, and d in each subarea, as shown in Figure 43(b), are measured. These correspond to the areas of background that should be covered with the object or the areas of the object that should be covered with background when the object is in an ideal position. They are obtained by simply counting the pixels in each subarea. Thus, the following values, p and q, are calculated:

$$p = a + b + c + d$$
$$q = a - b + c - d \qquad (3\text{-}10)$$

The values of p and q vary, as shown in Figure 43(c), as the object comes into the center of the window from the left and leaves the window to the right. The value of p is the total summation of errors that designates the degree of shape matching. Therefore, by checking that p is nearly equal to zero for all windows, we notice that the expected object arrives. We also notice the timing of object passage by checking that q becomes zero for one representative window or by checking that the averaged q for all windows becomes zero. A simple processing hardware can achieve these calculations by synchronizing with the TV camera scanning and can output the shape coincidence signal p and position-timing signal q at every vertical scan of the camera. In the bolting robot described before, these signals are appropriately delayed and are utilized the instant a detected bolt reaches a downstream position where a manipulator arm is located. These signals instruct the arm to start fastening if the object is a bolt, or to take a rear position to avoid collision if the object is a rib or tire.

3.3. Feature-Parameter Method

3.3.1. Utilizing General Parameters

The type, position, and orientation of an object can be recognized by analyzing the feature parameters of the two-dimensional figure that is the projected binary image of the 3D object. For stable thresholding, it is essential that the object has an adequate contrast to the background. Feature parameters that can be used include the area of the figure, its peripheral length, the number of holes involved, the total area of the holes, the moments of inertia about its principal axes, and the maximum and minimum distances from the center of gravity to the figure boundary. Combined parameters such as ratios and differences between these parameters can also be used.

For the objects $O_1, O_2, \ldots, O_m$ to be recognized, we can measure such parameters $P_1, P_2, \ldots, P_n$ as before in advance and can store them in a table, as shown in Figure 44. To measure these

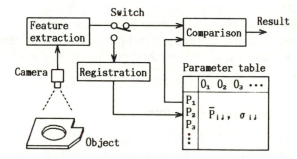

Figure 44. Feature-parameter method.

parameters, a teaching-by-showing method is used. In this method, an object O_i is fed into a field of vision of a TV camera at an arbitrary position and orientation and parameters P_j ($j = 1, 2, \ldots, n$) are automatically measured using binary-image analysis. Measurement is repeated a number of times for different positions and orientations so that the average and the deviation of each parameter are calculated and stored in a table. This is to cope with digitization error at the object boundary, which is due mainly to uneven illumination. In a recognition mode, measured parameters P_j for an unknown object are compared to those in the table, and the best-matched object is searched. That is, the object class O_i that satisfies the following equation for all j is judged as the class of the object:

$$| P_j - \overline{P}_{ij} | < k\sigma_{ij} \tag{3-11}$$

where k is a constant value, σ_{ij} is the deviation, and $\overline{P}_{ij}$ is the average. If there is a number of objects in an image, automatic segmentation by either the projection-distribution method or the labeling method can be applied beforehand.

In principle, this feature-parameter method can be applied to almost any object, if the object has a limited number of stable postures when it is placed on a table, and if a stable binary image is assured. In addition to obtaining the type of object, its position and orientation in the image plane can also be obtained using the center of gravity and direction of principal axes determined during parameter measurement. Therefore, positional information required to grasp the object can be easily output for physical classification of the object by a robot arm. This method, in which reference data are automatically formed by a teaching-by-showing techniques, is simple and easy to handle. Many general-purpose vision systems have been designed based on this method and are widely marketed for industrial use.

3.3.2. Utilizing Special Parameters

The parameters used in general-purpose industrial machine-vision systems are sometimes too numerous for certain applications, which may result in unsatisfactory performance and speed for object classification. Therefore, the parameters are sometimes restricted to a minimum number of more object-oriented parameters. There are many approaches along these lines, including detection of abnormal shapes in normally circular diode chips (see Reference 22). If the chips are cracked, or if aluminum deposition on their surfaces has come off, the chip images are deformed from their circular shape. Thus, in this application, area A of the bright aluminum surface and its contour length L are detected. A compound parameter L^2/A is calculated from these and then compared with 4π, which is the ideal value for a circle. The inspection machine for the diode chips is shown in Figure 45. In this machine, another method is jointly used, where the original image is compared with its counterpart rotated by 90 degrees about its center of gravity. The resulting numbers of different pixels can be used for the decision, thus increasing the reliability of abnormal shape detection.

The method using area A, contour length L, and their compound parameter L^2/A was also applied to shape classification of medicine tablets (see Reference 23). The machine is illustrated in Figure 46. The tablets are fed into hollows on the surface of a rotating drum and transferred to second and third drums, where the top and bottom surfaces of each tablet are inspected. If the tablet contains tiny cracks and/or flaws, it is ejected at the final

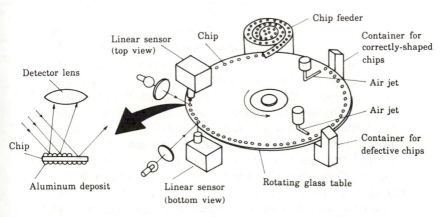

Figure 45. Shape classification of diode chips. (From Reference 8.)

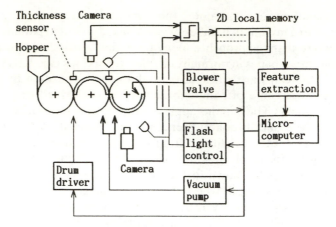

Figure 46. Shape classification of tablets.

position by an air jet. Measurement of area and contour length is executed by a processor hardware described earlier, consisting of a two-dimensional local memory for outputting 3×3 pixels and a logical operation and counting circuit (refer to Figure 29).

Automatic classification of natural products by their shapes is often useful as their packaging is sometimes concentrated to a short period of time. It also makes packaging for shipping easy when the products are classified into similarly shaped product groups. For example, as shown in Figure 47, cucumbers can be fed onto a belt conveyor and their silhouette images analyzed.[27] The median line of a cucumber is first extracted as a chain of center points between two edges in each scan line. The distance L between each end of the median line and the maximum distance W between the median line and the straight line connecting both ends are then found. The ratio W/L defines the degree of curvature. By using these parameters, L and W/L, the cucumbers can be automatically classified into several size and grade categories. The same techniques are also applied to classification of fish.[28] The silhouette images of fish automatically placed on a conveyor are analyzed to measure a number of feature parameters such as body length, tail length, body height, tail height, and ratios between

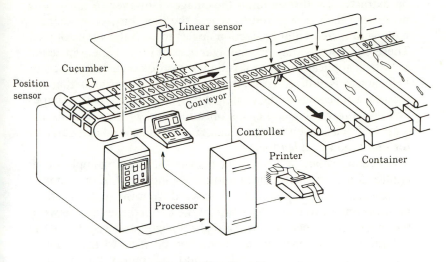

Figure 47. Classification of cucumbers.

them. An experimentally obtained discrimination function using these parameters can classify the fish into a number of types and sizes.

As in these examples, it is essential to choose appropriate feature parameters best suited to the objects to be recognized. This can also minimize the number of parameters used, thus making it possible to obtain high-speed classification.

3.4. Slit-Light Method

3.4.1. Depth-Difference Detection

The slit-light method is often used in industrial applications to detect an object's shape and position. A slit light is projected to an object and the reflection from the object is observed at a point distant from the point of projection. The reflected light forms a bright bent line in the observed image, whose shape and position depend on the shape and position of the object. The line can easily be extracted by thresholding the image. However, a fairly long processing time is required to repeat the light projection sequentially with incremental projection angles to scan the whole object. Therefore, usually only one slit light is used for judgment in actual industrial applications. The position of a crook detected in the slit-light image is utilized to control a manipulator arm so that the crook always comes to the center of the image field. As this simplified method can detect a crook once in every 17-ms vertical scanning field of a conventional TV camera, it can be used as a detector for real-time visual feedback.

A sealing robot, as shown in Figure 48, is an example of an application of this slit-light method.[29] As indicated in the detailed diagram, slit light is projected downward from the upper position to the seam of two plates. Because of the depth difference between the overlapping plates, the slit-light image observed from the oblique direction contains a crook. The position of the crook is detected by image analysis and kept at the center of the image by moving the manipulator hand. By this visual feedback, the hand position is always kept on the joint border, thus making it possible

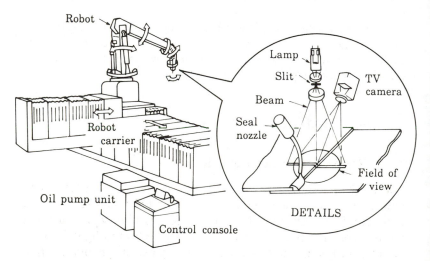

Figure 48. Sealing robot. (From Reference 8.)

to continuously feed sealant from a nozzle attached to the hand. This robot has been developed for sealing complex floors of car bodies. Once rough positions on the sealing route are taught, automatic sealing can be achieved by following the seam.

This method has also been applied to a welding robot, which automatically follows the welding line. The position of the seam is detected by the projection of a single slit light, and the following torch welds the seam line.

3.4.2. Profile Detection

In printed-circuit boards on which LSIs are mounted, the state of soldering at every lead is very important for circuit reliability. LSIs seem like skyscrapers, particularly in recent ultra-high-density printed-circuit board for mainframe computers and supercomputers, and inspection of leads is like looking down from the top of a building to a narrow street to check for defective paving tiles. A modified slit-light method, as shown in Figure 49, has been adopted for this inspection.[30] Instead of slit light, a laser spotlight

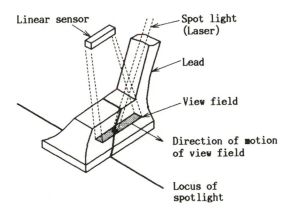

Figure 49. Inspection of LSI lead soldering by a modified slit-light method.

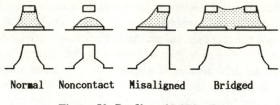

Figure 50. Profiles of LSI leads.

is scanned across the leads to measure their profile. To avoid erroneous signal detection by secondary reflection from neighboring LSI packages, the field of view is also scanned in synchronization with the spot scan. Connection abnormalities, such as noncontact, misalignment, and bridging, can be measured as a profile of spotlight locus, from whose shape the positions of abnormal leads can be found. The profile states of this method are schematically shown in Figure 50.

CHAPTER 4

Recognition of Defects

4.1. Automation of Visual Inspection

Mass-produced objects may sometimes contain defects such as cracks, or flaws, or have dirt or foreign material stuck to them. When object surfaces are intrinsically plain, it is relatively easy to detect defects. However, when they have complex normal patterns, it becomes very complicated. An abnormal pattern must first be discriminated from the normal patterns and then checked to see if it is allowable as a standard quality of the object. The allowable limit is usually decided from the aesthetic viewpoint or from potential for future malfunction. For example, an area of constricted wiring in printed-circuit boards does not pose any problems in an electrical-conductivity test, but concentration of current flow may cause long-term heat integration, resulting in an open circuit. Thus, to ensure reliability, it is important to remove such potential in advance by visual inspection.

However, it is usually difficult to provide quantitative decision criteria for future possibility using present patterns. Long-term engineering experience, based on accelerated life tests, is sometimes the only cue to deciding the boundary between what is normal and what is defective. As defects are usually diverse in type and shape, it is almost impossible to provide qualitative limits for all of them. Thus, to date, actual defect samples have been used to determine the limits. Human inspectors are asked to learn to recognize these sample patterns thoroughly before they engage in their tasks.

These human tasks pose the problem of inspection reliability as the inspection is not quantitative. Results are apt to deviate widely with individual inspectors, and likewise with their physical condition caused by fatigue and other factors. To remove such shortcomings and to maintain a constant standard of inspection, machine vision has been actively tried in defect-recognition applications. Another purpose has been to relieve humans from difficult, monotonous jobs. It has also been found, especially for recent VLSI devices, that it is almost impossible for humans to achieve consistent results. Therefore, expectations of automatic visual inspection are increasingly high.

4.2. Utilization of Pattern Features

4.2.1. Standard-Pattern Generation

To extract defective portions from a complex pattern using image analysis, it is important to first discriminate normal and abnormal patterns. When there are distinct morphological differences between them, they can be used as discrimination criteria. For example, abnormal patterns are usually smaller and sharper than normal patterns in printed-circuit boards for consumer electronic products. By using this feature, smaller portions can be detected as defect candidates (see Reference 9). To detect such small portions, the two-dimensional local memory described earlier can be utilized. Morphological operations such as AND and OR can be applied to the plural number of pixels extracted in parallel from the local memory (see Figure 19). Namely, a sequential execution of dilation, erosion, erosion and dilation outputs a pattern without small portions, and this pattern is compared with the original pattern by the exclusive-OR operation to find noncoincident pixels (see Figure 21). A defect-detection machine[31] (see also Reference 9) using this principle is shown in Figure 51, where one logical OR executes a dilation for a cross-shaped pixel area in a 3×3 matrix, and five ORs are arranged to cover a 5×5 pixel area. In this machine, dilation and erosion are executed in parallel, and the result is compared with the central pixel of the 5×5 matrix, which corresponds to the basepoint pixel in the original image. The two-dimensional local memory in this figure is illustrated as an equivalent spiral shift register into which image pixels from a TV camera are sequentially fed. Each pixel is thresholded to a binary form by using a floating-type thresholding circuit (see References 9 and 31), whose threshold value follows the image signal with a small time delay and

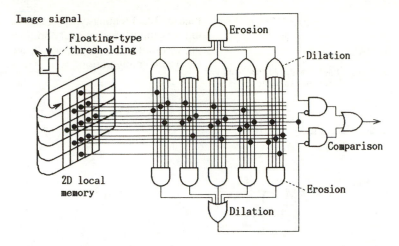

Figure 51. Defect-detection machine based on dilation and erosion method.

with a small gain reduction, as shown in Figure 52. Thus, the circuit can pick up small notches in both black and white levels in the image signal. When the output from the machine is connected to a monitor, only defective portions can be displayed. An overlapping display of the original image and the detected defect image in different colors is also possible. This machine also contains circuits to find the position coordinates and the number of defects and to store the results for later investigation and change. This method does not require storage of any standard pattern at all. Instead, the pseudostandard pattern is automatically generated from the input pattern in real time and used for self-comparison with the input pattern. The pseudostandard pattern is what the input pattern should be like if there were no defects. This machine was the first to utilize automatic inspection of complicated patterns based on image analysis.

4.2.2. Feature-Pattern Series

The density of printed-circuit boards used in industrial applications (or digital circuits) is usually higher than that of boards used for consumer products (or analog circuits). Industrial digital circuit

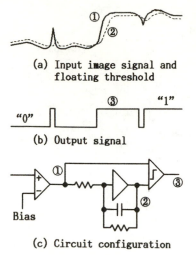

Figure 52. Floating-type thresholding. (a) Input image signal and floating threshold, (b) output signal, and (c) circuit configuration.

(a) Input image signal and floating threshold

(b) Output signal

(c) Circuit configuration

boards are already dense, and there is no longer any noticeable differences in morphology between normal and abnormal patterns. Therefore, the previous morphological operation, where detection of small portions implies the detection of defects, cannot be simply applied to present printed-circuit boards.

However, due to the limitation of the CAD system utilized for the design of the boards, the circuit pattern is usually configured as a combination of vertical lines, horizontal lines, and oblique 45-degree lines. Abnormal portions can be detected using this feature. For example, directional filters can be applied to eliminate patterns with normal directions, and remaining patterns can be regarded as defect candidates.

One such method utilizes a series of microscopic feature patterns. A 3 × 3 area, for example, is sequentially extracted and designated by a symbol according to its pattern arrangement. There are 2^9 ($= 512$) pattern arrangements in a 3 × 3 area. However, arrangements that can exist in a normal circuit pattern are limited to a small number of feature patterns, as in the examples shown in Figure 53. Here the possible patterns are designated by the symbols a, b, c, . . . , and the rest are designated collectively by the symbol e. Thus, this operation yields an image whose pixel

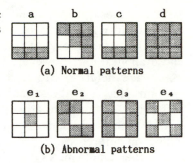

Figure 53. Examples of microscopic feature patterns: (a) normal patterns and (b) abnormal patterns.

(a) Normal patterns

(b) Abnormal patterns

data is replaced by feature symbols. Three consecutive pixels in both vertical and horizontal directions are then checked against this image. Some series of the feature symbols may be normal, such as *"aac"* in the horizontal direction, but other series must not exist in normal circuit patterns. When the normal feature-pattern series are obtained in advance, the series that do not belong to the normal series can be regarded as defect candidates.[32]

However, in recent high-density digital circuit boards for mainframe computers, the width of normal wiring is even smaller than the size of possible defects. Therefore, methods based on pattern features described here are difficult to apply in some cases. Consequently, methods based on the CAD design pattern have been developed and are being extensively used. These methods will be described later.

4.2.3. Use of Frequency Distribution

There are many types of objects produced in industry whose images are difficult to threshold into binary forms. Requests for inspection of these objects with gray levels have become increasingly high in recent years, as the availability of less costly high-performance image processors using image-processing LSIs increases.

One example is the inspection of plastic surfaces such as IC and LSI packages.[33] If there is a tiny hole in the plastic surface of such electronic parts, water vapor may enter the package in a humid environment. This can lead to a circuit malfunction, thus adversely affecting long-term reliability. To detect potential defects such as holes, hollows, and protrusions caused by inadequate filler quantity and air bubbles encountered in the molding process, automatic inspection can also be utilized. It is difficult in those applications, however, to obtain good images with highly contrasted defect portions because both normal and abnormal portions consist of the same materials and thus have no intrinsic differences. One method uses a spotlight to scan the object surface. Another uses planer scattering illumination to produce small reflection differences.

Another simple method for inspecting such plastic surfaces is to partition the observed image into several regions and to calculate a frequency distribution of brightness for each region. This partitioning can be effective in avoiding the effect of subtle shading due to uneven characteristics of the illumination and of the TV camera. The frequency distribution of brightness depends greatly on the particular defects in the region. When the maximum brightness f_{max}, minimum brightness f_{min}, and maximum-frequency brightness f_m are obtained, normal and abnormal objects can be discriminated, as shown in Figure 54, where $f_m - f_{min}$ and $f_{max} - f_{min}$ are used as axes for the two-dimensional feature space.

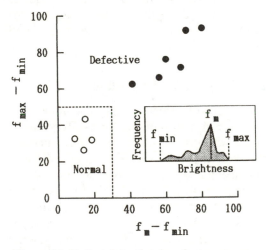

Figure 54. Defect detection in a feature space.

The frequency distribution of direction codes, instead of brightness itself, can also be used. A unique method of combining a direction-coded image and its frequency distribution is used to detect the number of persons waiting at an elevator, although this has no particular relation to the topic of defect inspection under discussion now. For a group-controlled elevator system, it is desirable for optimal service scheduling to always know how many persons are waiting on each floor. One such machine-vision system in which the input image is partitioned into several blocks is shown in Figure 55. The frequency of direction codes in each block is counted, and the histogram is regarded as a waveform. Its Fourier-transformed coefficients are considered to represent the block and are compared with the standard feature of the same type, which is stored in advance as background data. This standard feature is periodically updated automatically when no persons are observed, to cope with subtle changes due to camera drift and environmental illumination. The blocks showing differences are counted by adding appropriate weights to output an approximate number of persons in the scene.

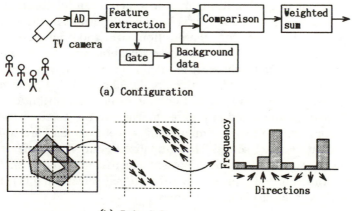

(a) Configuration

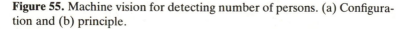

(b) Principle

Figure 55. Machine vision for detecting number of persons. (a) Configuration and (b) principle.

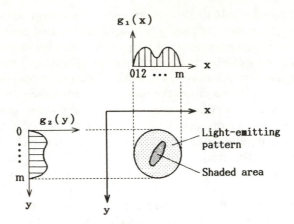

Figure 56. Projection distribution of light-emitting diode.

4.2.4 Use of Projection Distribution

Another example is the inspection of photodiodes, where the detection of subtle brightness differences is the key. Photodiodes usually have a circular light-emitting area, in which a careful observation sometimes reveals a subtle linelike shaded portion in the bright image. For this inspection, the position, size, and brightness level are first normalized. During this normalization process, abnormal diodes with a small light-emitting diameter and a low integrated intensity are rejected. Then, the circular light-emitting pattern is projected to both x and y axes to obtain projected distributions of brightness, as illustrated in Figure 56. The two projected distributions are regarded as one-dimensional waveforms, and are transformed into Fourier series. The normalized Fourier coefficients B_i are used as parameters to represent macroscopic shapes of the waveforms. For the y-projected x-directional waveform $g_1(x)$, the coefficients are calculated as follows:

$$B_i = 2 \sum_{x=0}^{m-1} g_1(x) \sin (\pi i x/m) \left/ \sum_{x=0}^{m-1} g_1(x) \right. \tag{4-1}$$

where m is the number of samplings of the waveform. The Fourier coefficients for the x-projected y-directional waveform $g_2(y)$ are obtained similarly. Judgment of normality can be performed in the parameter space using these coefficients as its axes. As normal diodes are not distributed uniformly in this space, a Mahalanobis distance normalized by the distribution of normal diodes is used as a measure for the decision. Namely,

$$D^2 = (X - \overline{X})^T K^{-1}(X - \overline{X}) \tag{4-2}$$

where $X = \begin{pmatrix} B_0 \\ \vdots \\ B_n \end{pmatrix}$ = feature vector of the inspecting pattern,

$\overline{X} = \begin{pmatrix} \overline{B}_0 \\ \vdots \\ \overline{B}_n \end{pmatrix}$ = feature vector of the averaged normal pattern, and

K = covariance matrix of normal patterns.

Distance D can be obtained for both x- and y-direction projection distributions. These are denoted as D_x and D_y, respectively. In a parameter space whose axes are D_x and D_y, normal diodes are concentrated in the portion near the origin, as shown in Figure 57.

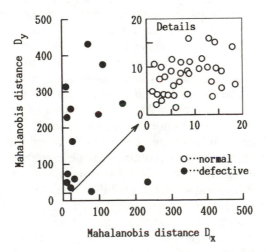

Figure 57. Defect detection by Mahalanobis distance.

Thus, it is possible to discriminate abnormal diodes from normal ones by checking if either D_x or D_y is abnormally large. The machine using this principle gives satisfactory results in photodiode inspection, where missing rates and false-alarm rates are comparable with those of human inspectors.[34]

As in the examples described, automated inspection for gray-level objects with a subtle brightness difference is incipient in industry. However, in general, the investigation of real-time image-analysis techniques for such gray-level objects is still at an early stage for vision applications. This is in contrast to binary-image analysis, and extensive efforts are expected in this in the future.

4.3. Pattern-Comparison Method

In this method, every portion of an object pattern to be inspected is sequentially compared to the equivalent portion of a normal object pattern. The principle of this method is to extract portions that are different, from two image signals: one from a camera looking at the normal object, and the other from a camera looking at the object to be inspected. Cameras with almost the same characteristics must be used to ensure stable imaging conditions of almost the same quality and to register the two object image signals as precisely as possible. If registration of the two patterns is insufficient, abnormal differences will be generated, especially at the pattern boundaries, thus signaling false alarms. This method has been utilized in many fields such as inspection of printed-circuit boards and shadow masks in color picture tubes.[35]

If the object to be inspected consists of a repetition of a single pattern, the adjacent two patterns can be compared with each other, thus eliminating the need for a normal standard pattern. This is possible because the probability of defects of the same size at the same position occurring in two adjacent patterns is usually very small. One example is a photomask used to form a pattern on a resist-covered silicon wafer in an IC fabrication process. If the photomask pattern is defective, the defect will be

transferred to the wafer. Thus, inspection of the photomask is
very important.

A pattern that corresponds to a certain layer of an IC chip is
repetitively formed on the photomask in both x and y directions.
The inspection machine usually moves the photomask continuously, and two linear sensors observe it from the vertical direction
through microscopes, as shown in Figure 58. The obtained signals
are continuously fed into buffer memories, and their output signals
are compared with each other in a defect-extraction circuit. The
output positions of the buffer memories are controlled to register
the positions of these signals. This can be called a two-chip comparison method as differences of two chips on the same photomask
are detected. However, the relative position between the two optical setups must be accurate, and the initial setting of the photomask must be precise, to ensure the complete positional align-
ment. This is not usually easy. To overcome these difficulties, a
method of feature comparison can be used instead of pattern comparison. This method is shown in Figure 59, where features are
extracted by using local-parallel processors consisting of two-
dimensional local memories and feature-extraction circuits. The
existence of these features is then compared within a predetermined allowable area. This method has been applied to LSI photomask inspection.[36,37]

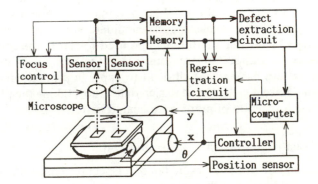

Figure 58. Pattern-based two-chip comparison method.

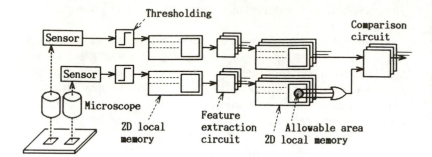

Figure 59. Feature-based two-chip comparison method.

4.4. Design-Pattern Referring Method

4.4.1. Effect of Design Pattern

If we proceed further with the idea of the pattern-comparison method just described, we reach the point where the design pattern is used as the normal pattern. Because recent complex electronic circuits are mostly designed using a computer, the design pattern in the CAD system can be diverted for this purpose. When the design pattern is referred, it becomes possible not only to detect defective portions as portions that do not exist in the design pattern, but also to adapt the processing at each position in response to the design pattern. This method also enables automatic judgment of what type of defect in what place each candidate is and its degree of seriousness. Thus, the method can be used to clarify the origin of the defect. It was stated previously that the method of using a feature pattern has become difficult to apply in printed-circuit inspection. However, the use of design data is now taking the place in automatic inspection of more complicated patterns.

4.4.2. List Matching of Connection Data

An example of printed-circuit board inspection based on design data first utilizes morphological operations to cut off the constricted portion of a wiring pattern and to connect the protruding

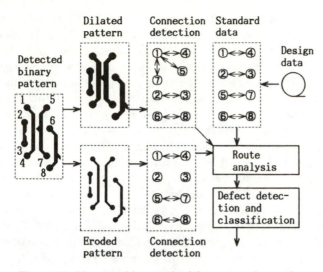

Figure 60. List-matching method for pattern inspection.

near-miss portions of two wiring patterns.[38] Then, the existence of connections between electrodes are checked. The erosion and dilation method is applied for these morphological operations. This method can, therefore, detect potential defects, and is illustrated in Figure 60. A logical OR filter and a logical AND filter of the same size can be applied to the conductor pattern for positive connection and disconnection. The size of the filter in this case is equal to the width of dilation or erosion, and this is set equal to the limit allowed in the inspection standard. Then, the connection table between the electrodes is obtained by image analysis, and the resulting list of connection data is compared, using a list-matching technique, with the list derived from the design data. The green sheets thus inspected are stacked in layers, heated, and completed as a printed board.

4.4.3. Pattern-Directed Processing

The examples described so far are single-layered, and intrinsically binary, patterns. However, in an IC wafer, a pattern is formed repetitively on previously formed patterns. Therefore, a multi-layered complex pattern is built up, forming a three-dimensional microscopic structure. Due to its uneven structure, the edges of each layer pattern can be easily seen, as if they were see-through patterns. This means that the image of the wafer contains very complicated gray levels and is impossible to simply threshold for binary-image analysis. Moreover, some layers, such as aluminum layers for wiring, contain grains or hillocks caused by depositing conditions and laid material. These may have a serious effect on recognition of defects if the inspection algorithm is not robust.

A pattern-inspection machine utilizing the design data has been developed[39,40] for IC patterns like these. The configuration of the machine is shown in Figure 61. Here, CAD data are stored in a memory, and from this, the design pattern is automatically generated. This generated pattern is compared with the image obtained by a linear sensor in pixel-to-pixel mode. Pattern positions are continuously measured by a position error-detecting circuit, and the resulting data are fed back to control the timing of design-pattern generation. Thus, the two patterns are made to coincide exactly. For this pattern-position measurement, logical AND-type correlation operations are used between the x- and y-directional

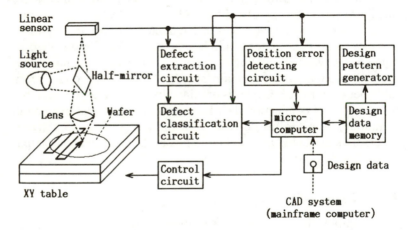

Figure 61. Configuration of wafer-inspection machine for logic ICs.

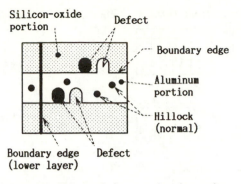

Figure 62. Schematic image of wafer surface.

edge patterns of the input pattern and the design pattern. The edge patterns are obtained independently, by the differentiation and thresholding technique for the input image and by logical filtering for the design pattern.

An image of the wafer is illustrated schematically in Figure 62. Each pattern area, such as an aluminum or silicon-oxide layer, has a respective brightness range. Therefore, pixels whose brightness deviates the normal range can be regarded as defect candidates. A defect-extraction circuit first extracts the defect candidates from the image by using the optimum threshold values selected in advance using the frequency distribution of brightness for each pattern area. The threshold values are automatically updated periodically by measuring the distribution. When the inspection is executed, information on which pattern area is presently scanned is obtained by referring to the design pattern. The threshold values and processing schema are thus switched to the ones that are most appropriate for the pattern area. This method can be called pattern-directed processing. In actual implementation, various processing schema are executed in parallel, and the results are switched to the appropriate one for that area. This process is shown schematically in Figure 63. The size filter based on the distance transform using propagation processing is utilized for this defect candidate extraction (see Figure 17) to eliminate the grain images.

For each defect candidate thus obtained, a degree of seriousness

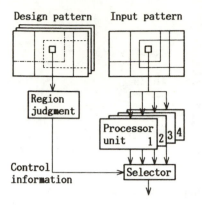

Design pattern   Input pattern

Figure 63. Pattern-directed processing.

is found. Only defects that are judged fatal are finally output from the defect-extraction circuit, which is a pipelined processor with real-time capability. As the degree of seriousness differs from layer to layer, the judgment parameters are switched using informa-tion on what area of the image is presently being scanned. This information is also obtained from the design pattern. Examples of decision criteria for fatal defects, applied to this machine, are shown in Figure 64 for the aluminum and silicon-oxide areas. They are almost the same as those used by human inspectors. Judgment of whether or not the decision criteria are met can be achieved by automatically measuring the pattern widths. This measurement is executed, as shown in Figure 65, through the distance transform of both detected defect pattern and design pattern. This is formulated as follows:

$$g_1(x,y) = \{g_1(x,y-1) + f(x,y)\} \wedge h(x,y)$$
$$g_2(x,y) = \{g_2(x,y-1) + h(x,y)\} \wedge h(x,y) \qquad (4\text{-}3)$$

where h is the design pattern, g_1 and g_2 are the output patterns, f is the input pattern (i.e., detected defect pattern), and $\wedge$ is the logi-cal multiplication (AND) between a binary pattern and (each bit of) a multilevel pattern.

As illustrated in Figure 65, the numbers of pixels are accumu-lated from one edge line to the other. That is, width information is represented at the boundary-edge portion. Therefore, only the

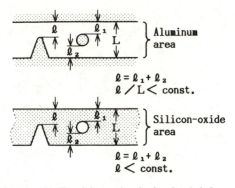

$$\ell = \ell_1 + \ell_2$$
$$\ell / L < \text{const.}$$

$$\ell = \ell_1 + \ell_2$$
$$\ell < \text{const.}$$

Figure 64. Decision criteria for fatal defects.

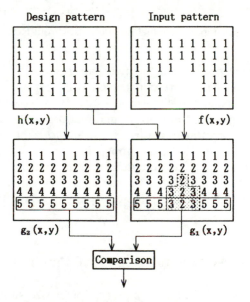

Figure 65. Pattern-width measurement.

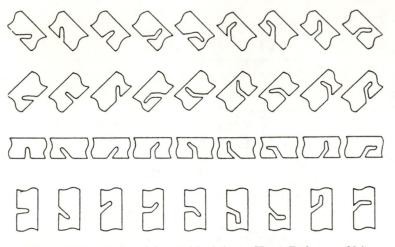

Figure 66. Examples of detectable defects. (From Reference 39.)

boundary-edge portions of the two patterns can be compared to judge how much they are constricted with respect to the design pattern. The distance transform in Eq. (4-3) is based on propagation processing in the top-to-bottom direction. In the machine, the distance transforms in four directions, left to right, upper left to lower right, top to bottom, and upper right to lower left, are implemented, thus enabling reliable detection of the degrees of seriousness for such complicated patterns as are shown in Figure 66. Examples of fatal defects detected by this method are shown in Figure 67.

4.5. Hybrid Method for Repetitive Patterns

For VLSIs with high-density patterns, pixel size should be minimized to enable the minimum detectable defect size to be small enough. Detection of submicron defects is becoming increasingly important, especially in leading-edge VLSIs such as memory devices. The heart of a memory device is the memory cell area, which contains many tiny cells. One machine used to inspect these areas combines the previous pattern-comparison method and the design-

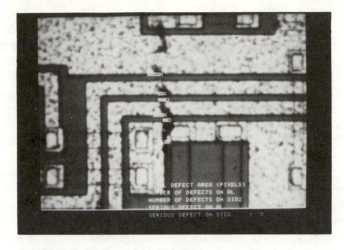

Figure 67. Original image and detected fatal defects. (From Reference 39.)

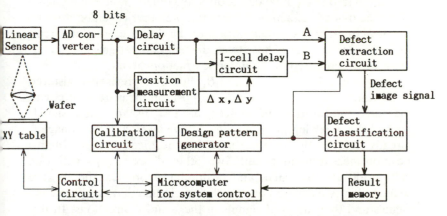

Figure 68. Configuration of wafer-inspection machine for LSI memories.

pattern referring method (see Reference 19). Its configuration is shown in Figure 68.

Because the memory cells are arranged two-dimensionally with constant pitches in x and y directions, adjacent memory cells are first compared to find defect candidates. However, two optical

setups for two adjacent memory cells are almost impossible, because the distance between them is too small and completely identical optical and imaging systems are not available. This machine
utilizes the monocular two-cell comparison method: the image is
compared with a delayed image with a period of one-cell pattern
length. Thus, images of neighboring cells obtained with the same
optical setup are always compared.

On the other hand, the pattern of a single cell and its repetitive
conditions, which indicate normal or mirror images, are stored as
the design pattern. This pattern can be automatically derived from
the CAD data. On inspection, the design pattern is repetitively
referred to as the image scanning proceeds.

In this machine, a wafer placed on a stage moves at a constant
speed in the x direction. A fixed linear sensor scans the wafer in
the y direction, thus obtaining a two-dimensional image. The analog image signal is converted to digital form with 8-bit brightness.
After passing through a delay circuit, the digital image signal is
separated into two paths: one is directly guided, as an image signal
A, to the defect-extraction circuit, and the other is guided through
a delay circuit with one cell period to the defect-extraction circuit
as an image signal B. Thus, two adjacent cell patterns, A and B,
are always compared in the defect-extraction circuit.

These two cell patterns, A and B, should be precisely registered
at all times. To do this, a real-time position-measurement circuit
measures the position differences, Δx and Δy, and these signals
control the delay length of the one-cell delay circuit. The image is
already delayed by the time used for the measurement. In this
position-measurement circuit, two other delay circuits with one
cell period are furnished, and positional errors, Δx and Δy, are
independently measured by correlation circuits that compare edge
signals and their delayed signals in the x and y directions. In this
case, the edge signals are obtained by spatial differentiation filters.

The configuration of the defect-extraction circuit is shown in
Figure 69. Here, the difference between two image signals A and B
is first taken and then thresholded by two threshold levels. A
binary image thresholded by the higher level is the image of nucleus areas of obviously different portions. A binary image thresholded by the lower level contains rather noisy patterns. However,

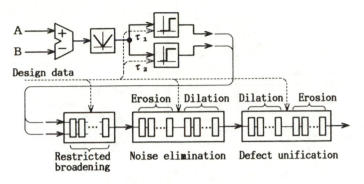

Figure 69. Defect-extraction circuit.

some of them are defects with their original shapes preserved. Therefore, the restricted broadening method described earlier (see Figure 22) can dilate the nucleus areas within the limit of the preserved shapes, thus outputing a "defect image" that only contains the reliable defect candidates. This circuit is realized as a pipeline structure, which is a serial combination of the previously described two-dimensional local memories and logical processing circuits. A process of eliminating isolated noises and a process of fusing neighboring small patterns, based on erosion and dilation operations, are also furnished by this circuit. The number of erosion and dilation operations is controlled by the design data, depending on the position of the image being scanned.

The defect-classification circuit is configured as shown in Figure 70. Here, feature parameters such as area and projected lengths are measured for each defect candidate involved in the inputting "defect image." The measurement is executed for each design pattern, which can be a compound type. For example, defects on layer 1, but not on layer 2, and at the same time on the boundary of layer 3, can be detected together with their area and size information. Such a compound design pattern can be produced by logical operations of design patterns of each layer.

To measure the feature parameters in one pass, a control image g is first generated from the inputting "defect image" f, as shown in Figure 71. In the image g, all holes and downward concave por-

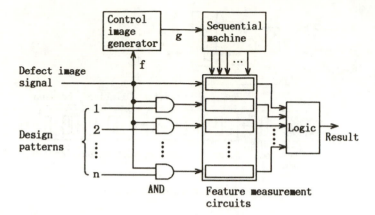

Figure 70. Defect-classification circuit.

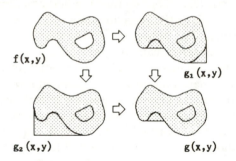

Figure 71. Generation of control image.

tions (bays) are filled. To do this, a recursive filter based on a propagation operation, as described earlier (see Figure 16), can be applied to generate two images, g_1, and g_2. These are the images for which the logical "1" pixel in the original image is propagated rightward down and leftward down, respectively. The logical AND of g_1 and g_2 gives the control image g. When we look at the two scan lines of this control image, only four types of pattern modes can be observed. Other modes do not appear because the control image no longer contains holes or downward concave portions.

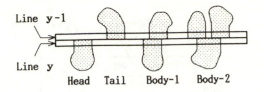

Figure 72. Possible pattern modes in two adjacent lines.

These pattern modes are "head," "tail," "body-1," and "body-2," as shown in Figure 72. For a "head" pattern, it signals the appearance of a new defect patttern at line y. Therefore, feature values of the pattern segment on line y, such as the segment length, must be stored as initial values. For a "body-1" pattern, a defect pattern up to line $y - 1$ is connected in a one-to-one correspondence with a pattern segment on line y. Therefore, the feature values up to line $y - 1$ must be stored after being updated by feature values of the pattern segment on line y. Accumulation of the segment length gives the area of a defect pattern. For a "body-2" pattern, multiple defect patterns are connected with each other. Therefore, feature values of these patterns up to line $y - 1$ must be united and then updated by feature values of the pattern segment on line y. For a "tail" pattern, feature values calculated up to line $y - 1$ must be released as the final result of the defect pattern because the defect pattern is terminated at the $y - 1$ line.

These functions can be realized by a sequential machine that is driven by the following 2-bit signal q:

$$q = 2^1 g(x,y-1) + 2^0 g(x,y)$$
$$= 0, 1, 2, \text{ or } 3 \tag{4-4}$$

The state-transition diagram of the sequential machine is shown in Figure 73. The control image is used only for judging the four types of pattern modes. The feature parameters are derived precisely from the original defect image. By this method, each defect can be evaluated with respect to the design patterns. A system-control microcomputer analyzes the final results, and statistical data are output for production control.

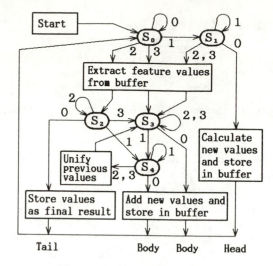

Figure 73. Sequential machine.

CHAPTER 5

Recognition of Surface Information

5.1. Mark Reading

Many types of surface information are given intentionally or unintentionally to objects. These include marks, characters, textures and colors, and sometimes become objectives of recognition. For example, product names, product sequence numbers, and production dates applying to each product or product class can be recognized and used for product control.

To classify objects flowing on a conveyor, the feature-parameter method described earlier can be utilized. This method is effective where the number of types of objects to be recognized is not extremely large, and the shapes of the objects are characteristic to each other, as in the case of parts handling in production processes. However, in physical-distribution systems, objects are usually stored in cardboard boxes, and the shape is thus restricted to a rectangular parallelepiped. Therefore, it is almost impossible to apply this size-and-shape-based method. Thus, recognition of surface information becomes important. Ornamental design patterns, product names, and product types printed on the cardboard boxes can be used as surface information. However, the difficulty arises from processing speed and hardware scale. It may be simpler to intentionally add a special mark for the recognition of each object so that it fits the ornamental design pattern.

In a typical physical-distribution system for factory automation, many types of products, usually produced in different factory buildings, are put into their own boxes. These boxes are placed on a single long conveyor that connects the buildings and are carried finally to a warehouse. At the entrance of the warehouse, a machine-vision system looks at the marks on the box surfaces, recognizes them, and assigns the destination of each object to the automatic routing and storing mechanisms. An example of such

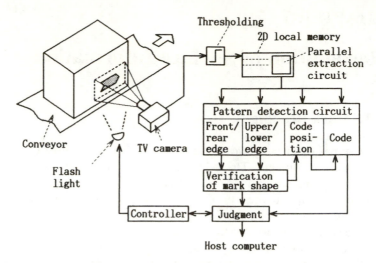

Figure 74. Machine vision for mark reading.

machine vision[41] is shown in Figure 74. The mark to be put on the object surface must be rather small but of high enough resolution for accurate reading. Also, the mark should be readable by a human operator in case of system failure. It may also be necessary to consider situations where the box is inclined at ±30 degrees to the conveyor surface and where it is rotated ±30 degrees on the conveyor surface. Even in those extreme cases, where the mark does not face the machine vision at right angles, the mark position must be extracted. For these purposes, a two-dimensional mark based on a 2 × 4 matrix configuration has been developed, as shown in Figure 75, where figures 1, 2, . . . , 9 are arranged. The figure 0 can be represented as an all-white 2 × 4 pattern. The last digit of the 9-digit figure can be used as a sum-check digit for higher recognition reliability. The mark is composed of arrowlike portions at both ends. The positions of these ends can be easily detected by a single pattern-matching circuit, as the patterns of these end portions, P_1 and P_4, are identical in shape, even though they are logically reversed. The matching process is illustrated in Figure 76. The upper and lower edges, P_2 and P_3, of the mark can also be

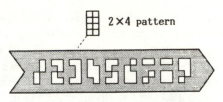

2×4 pattern

Figure 75. 2D mark for physical-distribution system.

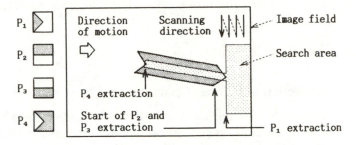

Figure 76. Principle of mark-matching process.

detected by a single matching circuit, as they are simply reversed. By determining the figure position as the center of those edge positions, the 2 × 4 subareas for each figure code are extracted. The decision of black-and-white subareas is determined by averaging the pixels that make up each subarea.

A conventional bar code can also be applied to physical-distribution systems. Characters from ordinary fonts can also be used. However, these should be robust for both positional and rotational errors encountered when the object is moving on the conveyor. The mark utilized here is robust in the sense that its existence is easily sensed by its characteristic shape even in very complicated background design patterns, and that, even with degraded printing quality, it can be recognized using pixel averaging. Furthermore, the two-dimensional mark used here can be made smaller than the intrinsically one-dimensional bar code, thus making it easy to fit into any background design of the cardboard boxes.

Besides factory automation, there is quite a number of physical-distribution systems that necessitate automatic classification, for example, classification of parcels in post offices, classification of freight in private overnight service companies, and classification of newspaper packages for regional shipping. However, relatively few successful results have been obtained in these areas. One reason is that object shapes are too diverse. An algorithm for reading surface information is not difficult to devise. The real difficulty is usually how the surface to be read can be automatically recognized and how the surface can be faced directly to the imaging device.

5.2. Character Reading

5.2.1. Inspection of Printed Characters

Characters printed on packages are one way of identifying the objects inside. There is no efficient way to discriminate electronic parts without characters on the packages. Thus, an inspection machine has been developed[42,43] to inspect printing quality and to feed back the result for controlling the printing mechanism. The configuration of the machine is shown in Figure 77. The printed surface of the electronic part is imaged by a TV camera, and stored in an image memory after thresholding. The binary image in the

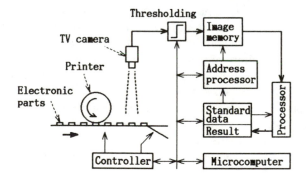

Figure 77. Inspection machine for printed characters.

memory is then analyzed by a microprogrammed processor. Because the characters and symbols printed on the surfaces are known, only standard data for specific characters and symbols can be used for comparison. The positioning of characters and symbols is very important because the parts are usually extremely small. Therefore, to absorb position-alignment errors, recognition is repeated several times for different shifted positions.

To judge the quality of printed characters and symbols, two methods are used jointly, as shown in Figure 78. A pattern-matching technique for partitioned subareas measures the difference in area in each partitioned subarea. If a subarea shows a large error, the printed pattern is regarded as inferior. The reason for

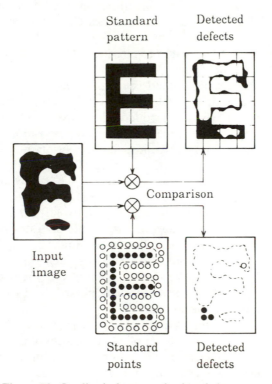

Figure 78. Quality judgment of printed characters.

partitioning is that a ±20% variation in average width is normal, but a 20% missing portion is obviously abnormal.

On the other hand, a point-matching technique utilizes essential points within and around the printed pattern. These points are specific to each character or symbol and correspond to places where the pattern cannot be omitted or added to. Thus, the number of differing points is an indication of print quality. With both methods, an inspection speed of 50 characters and symbols per second is easily achieved.

5.2.2. Industrial Character Recognition

The inspection machine for printed characters can be used as an optical character reader (OCR), without any modification, for various industrial applications. An unknown character can be recognized by finding a category giving the maximum degree of matching among the standard data prepared for all character categories. Where *a priori* knowledge can be used on what character must exist, an especially high-speed reliable recognition processing is achieved. This feature enables the machine also to be used for verification in a keyboard production process, to check whether appropriate keytops are placed in their correct positions.

This machine is also used for inspection of printed-circuit boards on which electronic parts are mounted. Before soldering, parts such as ICs and capacitors are checked to see if they are in their correct positions and orientations on the board. Though the parts are placed automatically on the board by an automatic parts feeder, there may still be an error when the operator presets the parts to the automatic feeder. This inspection is especially important in production of very expensive boards, such as for mainframe computers. The machine reads the characters on the package surface, and for capacitors, the direction of polarity is also checked by reading a symbol on the surface.

There are many other areas in industry in which the character recognition technique can be applied. This technological field can be called "industrial character recognition," in contrast to the conventional character recognition for document reading. Industrial character recognition does not necessitate a large repertoire of num-

bers and character fonts because these are usually restricted. Therefore, few problems remain unsolved as far as the algorithm is con-
cerned. However, contrasting the characters with their background
is sometimes difficult. Engine numbers on automobile engine
blocks are sometimes cast in relief. Automobile body numbers are
sometimes die-stamped. Automatic reading of these numbers is
important for production control. However, the character and background portions are of the same material. Therefore, contrasting
these two portions depends greatly on the external illumination
condition. There is no definitive solution for such applications yet.

When the characters to be read can be precisely placed in a
standard position, oblique illumination from an adjacent position
using fiber optics can be used. In general, plain illumination using
a light-scattering plate is effective for contrasting such characters.
An example of a machine for recognizing die-stamped characters
on cylindrical mechanical parts[44] is shown in Figure 79. Variation
in die pressure and wear of the die cause changes in depth and
width of stamped characters. Therefore, thresholding is controlled
to yield binary character patterns with nearly constant line widths.
The relation between the widths and threshold levels is stored in
advance in a look-up table, and the levels are determined from the
measured widths by referring to this table. To measure width W,
two feature parameters are first found: one is the number of pixels
comprising the edge of the character (i.e., boundary length L),
and the other is the number of pixels that are surrounded by the

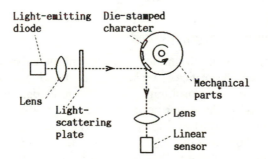

Figure 79. Machine vision for die-stamped character recognition.

edge (i.e., area A). Width W is obtained from the ratio $W = 2A/L$. The recognition can be executed by the pattern-matching method with standard patterns, after normalizing the character position at its center of gravity.

5.3. Recognition of Texture

Texture analysis is a basic technique in three-dimensional scene analysis for deriving regions of object surfaces and their varying depths from a two-dimensional image. However, this technique has not yet been exploited in industry. Although object classification based on texture analysis is conceivable, there is no specific need at present.

However, there are definitely needs in qualitative feature recognition, including detection of defective texture portions due to registration error and detection of defects buried in complex normal textures. Inspection of fabrics is a typical example. However, no cost-effective solution has yet been found.

Smoothness of processed object surfaces are sometimes judged visually or tactually. However, independent inspection machines for automatic surface inspection are rather inefficient. Processing machines must be equipped with an inspection function to control the surface finish.

5.4. Recognition of Colors

5.4.1. Color-Filtering Method

Few applications of automatic color recognition exist yet in industry. However, the principal technology is already well developed. The only outstanding problem is to eliminate recognition errors by stabilizing the color temperature of the illuminating light source and by avoiding short and long-term signal drift in the color sensors. Figure 80 shows two typical configurations of color-detection methods based on color filtering. A point light sensor such as a photomultiplier captures light projected from a white light source

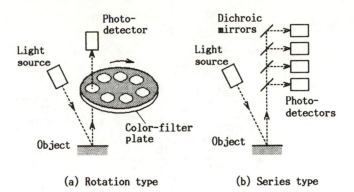

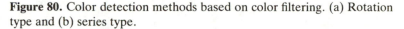

(a) Rotation type   (b) Series type

Figure 80. Color detection methods based on color filtering. (a) Rotation type and (b) series type.

and scattered at the object surface through n narrow band-pass color filters. The outputs A_i ($i = 1, 2, \ldots, n$) for these color bands form a vector that features the object surface, and the vector is regarded as a point in n-dimensional feature space whose coordinates are A_i. The color category is found from the distance between the point and the vector points of standard colors. When $n = 10$, it is usually possible to find fifty different colors.

The automatic color-recognition method based on color filtering, which is shown in Figure 80(b), is applied to an automatic digitizer for color drawings.[45,46] Here, the drawings are sampled at a pitch of 0.1 millimeter in both x and y directions. The pixel data are sequentially fed into a color-recognition circuit, shown in Figure 81, and are converted to color codes in real time. The output is, therefore, the image whose pixel data are color codes, for example, 1 for red, 2 for blue, 3 for black, etc.

Four color interference filters having center wavelengths of 455 (blue), 515 (green), 675 (red), and 765 (infrared) nanometers, each with 25 nanometer full-width half-maximum (FWHM), are used in this circuit. Because drawings made with colored pencils usually contain blurs due to drawing pressure variations, the recognized color categories are restricted. In this circuit, six to eight colors can be reliably recognized. The principle of recognition is as follows.

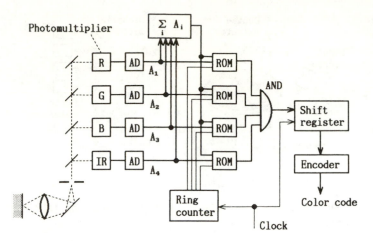

Figure 81. Example of real-time color-recognition circuit.

Let us denote A_i as the output from the ith sensor. The total sum ΣA_i for all i is calculated by an adder circuit. The following equation is then calculated:

$$| (A_j / \sum_i A_i) - S_{ij} | < T_{ij}$$

$$(5\text{-}1)$$

where S_{ij} denotes the standard value of $A_j / \Sigma A_i$ for the jth color and T_{ij} is the allowable tolerance level determined in advance. This calculation is executed at high speed by a table look-up method, using ROMs (read-only memories) addressed by A_i and ΣA_i. In the ROMs, the tolerance levels T_{ij} for each color j are stored. The levels are sequentially selected by a ring counter in response to the color selection, and logical "1" is output when Eq. (5-1) is satisfied. When outputs for all i are "1," the color is determined as j. A shift register is provided to store this ANDed output and is shifted in synchronization with the ring counter. The content in the shift register at the end of a cycle is converted to the color code.

In this method, a color is represented as a point in n-dimensional feature space, and the distances to the standard points are evaluated with some allowable tolerance. That is, the colors are measured relative to each other. Therefore, the color filters do not necessarily have single-peak narrow band-pass characteristics. In principle, any combination of filters can be used, provided they can cover a wide range of color wavelengths.

5.4.2. TV Signal-Processing Method

The three principal colors in a color TV camera, R, G, and B, can be converted to r, g, and b by the following calculations:

$$r = R/(R + G + B)$$
$$g = G/(R + G + B)$$
$$b = B/(R + G + B) = 1 - r - g \qquad (5\text{-}2)$$

A color can be represented as a point in a plane whose axes are r and g. In application to recognition of object surface colors, the r–g plane can be partitioned into regions for each color category in advance and can be used to judge the region in which the measured point falls. This method can discriminate from fourteen colors when $\pm 20\%$ illumination variation exists and up to twenty colors when there is no obvious brightness change.[47]

For human perception of colors, hue, saturation, and luminosity are important parameters. Hue depends on light wavelength, saturation relates to color purity, and luminosity represents brightness of the color. In industrial uses, there are a number of possible applications in which objects are automatically classified by recognition of surface color or color-coded marks added to the object surfaces. In these applications, detection of hue and degree of saturation is important.

Signals from a conventional color TV camera consist of a luminance signal Y and chrominance signals $R - Y$ and $B - Y$. In the two-dimensional plane with the axes of these two chrominance signals, achromatic colors that do not have hue and saturation components are concentrated to a point (b_0, r_0). Therefore, an arbitrary color is represented as a vector with displacement s and angular deviation θ in this plane, as shown in Figure 82(a). Namely,

$$\theta = \tan^{-1}(r_1/b_1)$$
$$s^2 = r_1^2 + b_1^2 \qquad (5\text{-}3)$$

where $r_1 = (R - Y) - r_0$ and $b_1 = (B - Y) - b_0$.

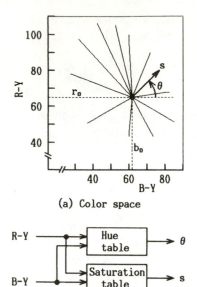

(a) Color space

Figure 82. Color representation in color space. (a) Color space and (b) color conversion.

R-Y ⟶ Hue table ⟶ θ

B-Y ⟶ Saturation table ⟶ s

(b) Color conversion

For color recognition, this plane can be partitioned radially from the center (b_0, r_0), as shown in Figure 82(a). Hue θ and saturation s are calculated in advance for all combinations of $R - Y$ and $B - Y$, and stored in ROMs as tables. Thus, values θ and s are output at high speed by two-dimensional addressing of the ROMS with $R - Y$ and $B - Y$, as shown in Figure 82(b). Furthermore, the color-code number can be stored in the ROMs so that the color code is output directly in real time by $R - Y$ and $B - Y$ addressing. This method can be easily applied to recognition of up to 12 colors when a ±20% illumination intensity change is allowed.

Color recognition can be applied to verification of resistors on a printed-circuit board to ensure that correct parts are at their appropriate positions. Color categories for resistor color codes usually involve a certain variation in spectral characteristics, depending on the resistor manufacturer. Therefore, control of illumination and stabilization of image signals are the keys to reliability.

CHAPTER 6

Image Processors
and Future Machine Vision

6.1. Configurations of Image Processors

There are basically three types of image processors. The first is the fully parallel type, whose processing elements (PEs) are assigned to each pixel with a one-to-one correspondence. The second is the local-parallel type, in which a plural number of pixels is locally extracted in parallel and processed as a pipeline. The third is the multiprocessor type, whose processing units (PUs), each of which is a combination of a processing element (PE) and local memory, execute image processing in parallel.

6.1.1. Fully Parallel Image Processor

The fully parallel image processor can usually be configured as shown in Figure 83, so that each processing element (PE) can receive data from its adjacent eight PEs. This particularly fits the 3 × 3-size filtering operation. However, for a larger filter, controlling data transfer becomes very complicated. A rather simple way to execute larger filtering is to decompose the filter into a number of 3 × 3 filters. These decomposed filters can then be executed sequentially.

The difficulty in controlling data transfer leads to difficulty in programming. Thus, it is seldom used in industrial applications where simplicity is the highest priority. Furthermore, it necessitates massive processors of 64K, even for the 256 × 256 image, which is one of the most popular image sizes for industrial applications. It also poses a connection problem between numerous PEs. To solve this problem, the processor can be restricted to, say, 1-bit processing, and a number of PEs, for example, 8 × 8 (= 64) PEs, can be integrated into one LSI chip. However, even in this case, 32

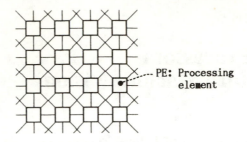

Figure 83. Basic structure of fully parallel image processor.

$\times 32(= 1024)$ LSIs must be implemented for the 256×256 image, still making industrial applications difficult. This fully parallel-type image processor originated from the idea of simulating neural networks. Novel architectures may be found in the future, in response to the progress of recent active research in neural networks.

6.1.2. Local-Parallel Image Processor

Local-parallel image processors are, as stated before, widely applied in industrial automation. They basically consist of a parallel extraction circuit of pixels of a certain size from a local area and a processing circuit of the extracted pixels. For the parallel extraction circuit, a two-dimensional local memory described earlier can be effectively used (see Figure 14). The local-parallel image processor is based on the principle that the input one-dimensional image signal is once converted into local-parallel two-dimensional information, then processed, and again returned to a one-dimensional signal. That is, it is suited to pipeline processing to which an original image signal is fed and from which a processed image signal is extracted. Thus, by regarding the combination of a parallel extraction circuit and a processing circuit as a processing element (PE), a number of PEs can be connected in series, as shown in Figure 84(a). By assigning various processing functions to each PE, real-time high-speed processing can be achieved. The wafer-inspection machine previously described utilizes this special-purpose pipeline processor, and continuously achieves 10 GOPS (giga operations

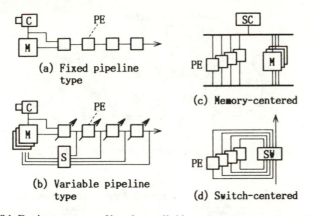

Figure 84. Basic structure of local-parallel image processor, where PE is the processing element, M is the memory, C is the camera, SW is the switch, S is the selector, and SC is the system controller. (a) Fixed pipeline, (b) variable pipeline, (c) memory-centered, and (d) switch-centered.

per second) equivalent speed when it is assumed to be processed by a conventional von Neumann type computer.

There are many other modified versions of pipelined local-parallel image processors. Their PEs can, of course, be variably structured, as shown in Figure 84(b); or arranged as memory-centered, as shown in Figure 84(c), or configured as switch-centered, as shown in Figure 84(d). A unique structure is the token-ring type, shown in Figure 85, where a data-flow-type control scheme is adopted instead of local-parallel-type image extraction.[48]

There is another conventional method for extracting local pixel data in parallel other than the two-dimensional local memory. Here, the bank-structured image memory is used, as shown in Figure 86. Adjacent pixels are stored in different memory banks and read out in parallel via only one memory access. The output pixel data can be rearranged to have a normal arrangement (see Reference 43). Various programmable PEs can be applied to the output pixel data, enabling high-speed processing.

The structure of the local-parallel-type image processor is rather simpler than that of other types. Therefore, most general-purpose image processors marketed are of this type. Typical examples of

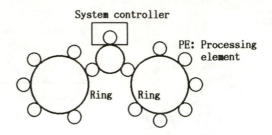

Figure 85. Token-ring image processor, TIP (NEC).

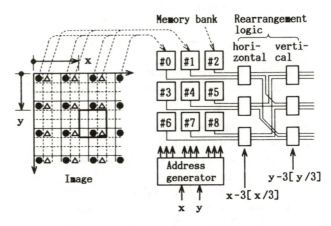

Figure 86. Image memory for parallel pixel output.

general-purpose image processors are shown in Figures 87 and 88, which are both memory-centered types. In these processors, PEs are arranged between input and output buses, and image data are transmitted between memories by controlling the PE paths. An example of the switch-centered general-purpose image processor is shown in Figure 89, where the paths to and from the PEs are switched. These processors are rather flexible and can function as general-purpose processors in contrast to the special-purpose pipeline processor with PEs connected in series. However, processing speed is slower than that of pipeline processors specialized and optimized for the specific problem.

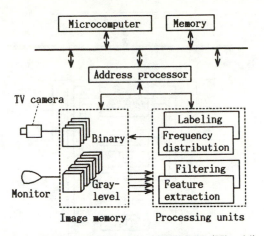

Figure 87. Image processor, HIDIC-IP (Hitachi).

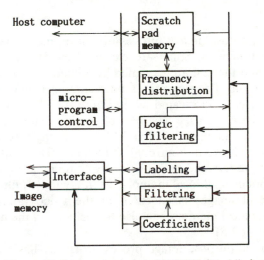

Figure 88. Image processor, TOSPIX (Toshiba).

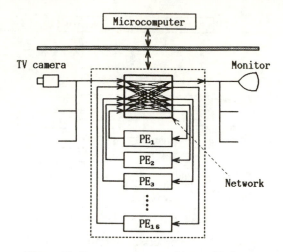

Figure 89. Image processor, Idaten (Fujitsu).

General-purpose image processors have many basic image-processing functions. Some are executed by PEs as a hardware operation and the rest are executed by a main processor as a software operation. The operations necessitating many memory accesses are usually fabricated as special-purpose image-processing LSIs, and these are used as a part of the PE. Filtering is such an operation, and an example of an LSI for filtering (see References 16 and 17) is shown in Figure 90. It includes 1×4 shift registers, and several adders and multipliers. By combining these LSIs and the two-dimensional local memory, as in Figure 91, filtering for 8-bit gray-level images is executed at the video rate. The two-dimensional local memory can also be included in the new version of the filtering LSI.[49]

The general-purpose image processor usually has a large number of functions and can be utilized for machine vision in industrial applications. However, its cost effectiveness is still a problem in some applications. Therefore, much simpler image processors with fewer functions, yet programmable and therefore general-purpose, are being developed. An example of such an industrial image processor[50,51] is shown in Figure 92, which has the capability of process-

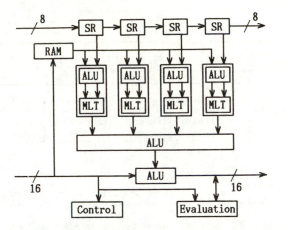

Figure 90. An LSI for image filtering, ISP, where RAM is the random-access memory, ALU is the arithmetic logic unit, MLT is the multiplier, and SR is the shift register.

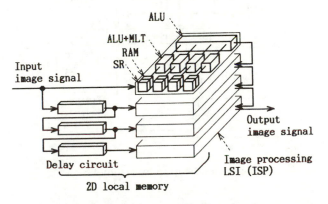

Figure 91. Video-rate filtering circuit.

ing gray-level images at high speed. Therefore, they are finding applications in many industrial fields as key components of automatic production machines. LSIs used for this processor are filtering LSIs described before, feature-extracting LSIs shown in Figure 93, and some others. The feature-extracting LSI can execute func-

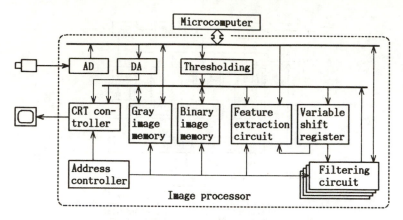

Figure 92. Industrial image processor, SBIP (Hitachi).

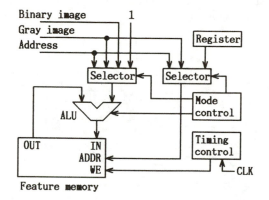

Figure 93. Feature-extracting LSI.

tions such as frequency-distribution calculations and projection-distribution calculations by switching the functions with a control signal given from outside (see Reference 50).

6.1.3. Multiprocessor Image Processor

The multiprocessor image processor can execute a variety of processings as a general-purpose image processor. The typical configu-

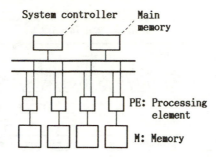

Figure 94. Basic structure of multiprocessor image processor.

ration is shown in Figure 94, where each processing unit PU (= PE + memory) executes image processing in parallel. An image can be partitioned into subregions and these can be assigned to different PUs. Different processings for a single image can also be assigned to different PUs, and different processings for different images can be assigned to different PUs.

The multiprocessor image processor usually has complex and large-scale hardware and is difficult to program. Therefore, it has been restricted to laboratory use. However, the realization of PUs as VLSI chips has increased the possibility of applying the multiprocessor image processor to industrial use.

Each PU is a multipurpose module and is not for a specific function. All possible functions are designed into the PU so that it can achieve every necessary operation independently of other PUs. Because conventional chips, such as microprocessor chips, are not suited to image processing, a novel architecture is usually attempted. High-speed performance is one requirement, not only for the PU itself, but also for the set of PUs integrated as a multiprocessor. Figure 95 shows an example of the processing-element LSI that can be used as a part of the PU. This PE LSI has signal-processor architecture, in which an independent addressing circuit, a two-layered command control circuit, and a bit-processing circuit are implemented for 20-MOPS (mega operations per second) image processing (see Reference 18).

The multiprocessor image processor using 64 LSIs as PEs[52,53] is shown in Figure 96. In this image processor, PUs, each consisting

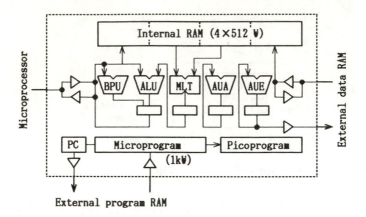

Figure 95. Processing element, DSP-i, where BPU is the bit-processing unit, ALU is the arithmetic logic unit, MLT is the multiplier, AUA is the addressing unit A, AUE is the addressing unit E, and PC is the program counter.

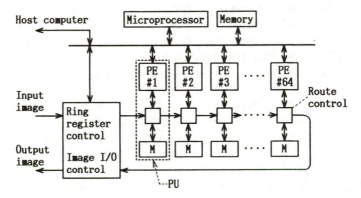

Figure 96. Multiprocessor image processor, GPIP (Hitachi).

of a combination of PE and attached memory, are connected by a shift-ring bus. A shift-register portion of the shift-ring bus can be configured as shown in Figure 97, and it controls the data transfer between the PE, the memory, and the bus, in responding to flags from the controller. For example, a large-scale image can be parti-

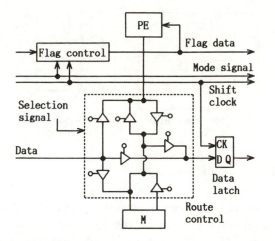

Figure 97. Data paths in ring-register interconnection.

tioned, as shown in Figure 98, in a partially overlapping mode, and each subimage is assigned to a PU. The PUs process their own assigned area and send out the resulting data from the overlapped area to the shift-ring bus simultaneously. The bus shifts these data altogether to the adjacent PUs. Thus, the process ensures correct processing results at the partitioned boundary even for two-dimensional filtering operations. Furthermore, overheads common to all multiprocessor-type architectures become relatively small in this case. Although the 64 PUs are connected one-dimensionally on the bus, each PU can be accessed by a two-dimensional address (i,j), where the address $(0,0)$ in particular denotes the assignment of all PUs for broadcasting the data and programs. Thus, the SISD-, SIMD-, MISD-, and MIMD-type operations can be executed (where S denotes "single," M denotes "multiple," I denotes "instruction", and D denotes "data"). The average total processing speed is 1 GOPS.

Each PU can handle images of up to 512 × 512 pixels, thus resulting in the image size of 4096 × 4096 as a total image processor. Each pixel consists of 16 bits, and each bit can be reconfigured as an independent binary image. Thus, a binary image of 16,384 × 16,384 pixels can be processed by this 64-PU configuration. Be-

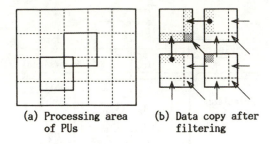

(a) Processing area   (b) Data copy after
    of PUs   filtering

Figure 98. Filtering for a large-scale image. (a) Processing area of PUs, and (b) data copy after filtering.

cause the maximum number of PUs that can be installed is 4096, the full configuration can handle the image of 262,144 × 262,144 gray-level pixels or 1,048,576 × 1,048,576 binary pixels.

6.2. State of Image Processing

6.2.1. Required Performance

As described in the previous sections, image processing is a complicated process requiring many operations on a vast amount of pixel data in an image. Therefore, required processing time is liable to be considerably long. However, allowable processing time is limited by constraints other than the image processing itself. For example, when image processing is to be applied to a certain assembly task, it must be finished within the limit of the cycle time of the assembly machine. When applied to a certain product-inspection task, it must be as fast or faster than a human inspector. It is also expected to achieve better than human performance.

The required image-processing times in various application fields[54,55] are shown in Figure 99. For example, the processing time required for robot vision is usually in the order of 0.1 through 1 second and is a maximum of a few seconds. In semiconductor inspection, processing time must also be in the range of 0.1 to 1 second for individualized chips, or 10^2 to 10^3 seconds (a few minutes to a few tens of minutes) for wafers. For these applications,

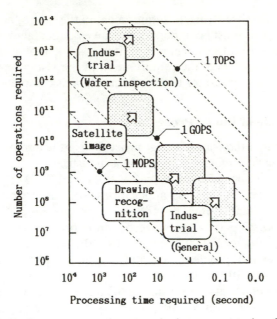

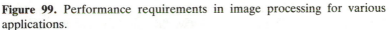

Figure 99. Performance requirements in image processing for various applications.

many operations are usually required. The number of operations is estimated at 10^7 to 10^8 for robotic vision and 10^{12} to 10^{13} for wafer inspection. The required processing speed is the ratio of required number of operations to required processing time. Therefore, the equispeed lines are expressed by the oblique lines in Figure 99. By comparing each application requirement to these lines, the required speed for image processing can be understood to be fairly high. For example, the processing speed of existing computers is approximately within the range 0.1 to 5 MOPS (mega operations per second) for microcomputers, 10 to 100 MOPS for mainframe computers, and 500 to 2000 MOPS for supercomputers. Therefore, for a certain usage, it is necessary to realize an ultra-high-speed image processor with a speed far exceeding that of the supercomputer at less than one hundredth, or even one thousandth, the cost. Even if this requirement is met, it will inevitably be faced with an even more stringent requirement in the future, as shown in Figure 99. This must be understood as an essential factor in image-processing applications. Therefore, it is possible that a production machine will not meet the requirement, as the vision requirement could go one step further before the machine is completed.

6.2.2. Scale of Image Processing

Here, consider the scale of image processing as the number of pixels constituting the image. Conventional industrial image-processing applications utilize 256×256 ($= 6 \times 10^4$) or 512×512 ($= 3 \times 10^5$) pixels. In office automation applications, such as in workstations, images with 1000×1000 ($= 10^6$) pixels are common. In satellite image processing or drawing-recognition processing, $10,000 \times 10,000$ ($= 10^8$) to $30,000 \times 30,000$ ($= 10^9$) pixels are used. The wafer-inspection problem described earlier requires $600,000 \times 600,000$ ($= 4 \times 10^{11}$) pixels when a 5-in. wafer is fully inspected for submicron defects at a resolution of 0.2 micrometers square. Thus, image data handled in image processing are becoming enormous (see Reference 55), as shown in Figure 100. By comparing other familiar numerical figures, such as world population or the number of neurons in the brain, the scale can be well understood.

The vertical axis represents the time required for image processing when a 20-MOPS-class mainframe computer is assumed to be used. Relatively simple image processing for industrial use usually takes from a few seconds to more than 10 seconds. For actual applications, this should be reduced to 0.1 to 1 second, or sometimes in the order of 0.01 second. Thresholding the image in the early stages of image processing, thereby replacing the necessary processing with logical operations, was used to speed up processing in past industrial applications. In wafer inspection, a general-purpose computer requires one month, operating 24 hours per day, to inspect a single wafer. This is obviously impractical. As described earlier, a special-purpose image processor with pipeline architecture reduces this to less than 1 hour. Thus, the revolution in image-processor architecture and the realization of its key portions by LSI technology are extremely effective in high-speed im-

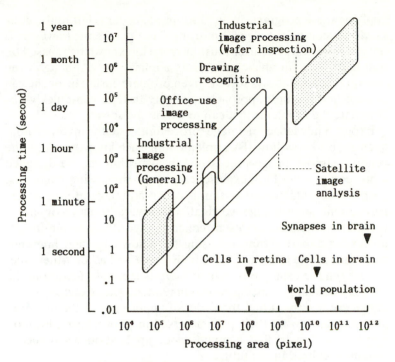

Figure 100. Scale of image processing.

age processing, although they are, of course, not infallible. More intensive research is needed along these lines.

6.3. Future of Machine Vision

6.3.1. Image-Processing Expert System

Various basic programs are prepared in the general-purpose image processors currently on the market, and these are usually executed as commands by a simple procedure. When the image processor is intended to be used for an industrial application, the method of applying these programs in an appropriate order is left to the user. Devising an efficient algorithm using these programs, however, is not a simple task. The user must have a wide knowledge, based on past experience in image-processing applications. Although even an expert cannot always solve a given problem easily, he or she can at least find the best approach relatively fast, or can quickly discriminate the problems that can or cannot be solved.

Expert knowledge in image processing usually involves non-quantitative know-how. By implementing such knowledge into the general-purpose image processor, an algorithm can be designed more easily, even by inexperienced image-processing engineers. Such a knowledge-based system is called an image-processing expert system and is being investigated by several research organizations. It is intended to design concrete algorithms in the image-processing expert system by using the knowledge and inference mechanism. The knowledge is usually expressed as an if-then rule, and its generic example is that; if there is a certain feature in the image, a certain processing is recommended. The objectives of the design are mainly to find the order of application of the prepared programs and to fix their parameters. Image-processing technology is expected to find a much wider application when such a system is realized in the future.

The nature of the expert system is intrinsically a conversational type. For example, when the user designates a region in a displayed image, the system responds to it by displaying the histogram of brightness in the region, recommends a method of thresholding, and presents the resulting binary image. Figure 101 shows an example of the basic configuration of the expert system that has been conceived and experimentally developed. One of the final goals of this system is to automatically generate an execution program. However, the system is still basic and elementary.

It seems hasty, however, to assume that the algorithm can be easily applied. The expert system can be useless if firm verification and appropriate modification of the obtained algorithm are omitted. The expert in image-processing applications, in fact, devotes considerable time to verifying and modifying the algorithm. When a new application problem is given, the expert usually starts by investigating whether the problem can be solved without using

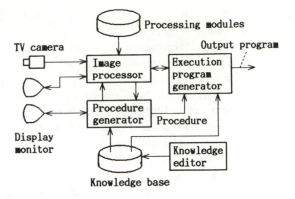

Figure 101. Configuration of image-processing expert system.

image processing. Factors that make the problem difficult are first considered, and the possibility of easier solution, such as by mechanical position registration, orientation normalization, and, in some cases, redesign of objects, is examined, together with cost effectiveness in solving the problem and running the system that will be developed. Real experts are those who understand the present difficulties in image-processing applications.

6.3.2. Knowledge-Based Processing

In typical industrial applications, a recognition algorithm is first designed with respect to the objects to be recognized and the objectives to be achieved. The image processor is then furnished with a program based on this algorithm. Design of the algorithm is usually performed by an image-processing expert, and the detailed procedure and parameters to be used are determined. The algorithm usually starts with low-level processing, which directly accesses the pixel data, and gradually goes to higher-level processing. That is, the algorithm is hierarchical, and bottom–up processing is usually executed. The algorithm is the so-called procedural knowledge for image processing and is implemented in the image processor as a program by a designer.

Conversely, a nonhierarchical image-processing method has

been proposed.[56] This method determines, for example, the edges of an object in an image by an easily detectable order. Then, the existence of other edges is assumed from the object information obtained. A model-based method[57] utilizes structural models of objects, where a most-probable model is selected from the information obtained until then by low-level processing, and the features to be recognized next are determined for subsequent processing. By using the object knowledge, and by adding the top–down processing as shown in these examples, more effective image processing can be expected than by bottom–up processing only.

In these processings, not all of the procedures are defined in advance, but are changed by intermediate decisions, according to the state of the image. By representing the knowledge explicitly as a declarative form, the possibility of simplifying the design of the algorithm increases. In this case, not only the object knowledge, but also knowledge of the methodology of image processing can be implemented, together with the inference mechanism, thus resulting in effective image processing, combining top–down and bottom–up processing. At present, however, there is no efficient and cost-effective hardware available for knowledge-based processing of images. Therefore, rather slow image processing is likely to become even slower. Although steady progress can be expected, it will be some time before this method becomes available for industrial applications.

6.3.3. Future Development

Although image processing has been successfully applied to many industrial applications, there are still many definitive differences and gaps between machine vision and human vision. Past successful applications have not always been attained easily. Many difficult problems have been solved one by one, sometimes by simplifying the background and redesigning the objects. Machine-vision requirements are sure to increase in the future, as the ultimate goal of machine-vision research is obviously to approach the capability of the human eye. Although it seems extremely difficult to attain, it remains a challenge to achieve highly functional vision systems.

The narrow dynamic range of detectable brightness is the biggest cause of difficulty in image processing. A novel sensor with a wide detection range will drastically change the aspect of image processing. As microelectronics technology progresses, three-dimensional compound-sensor LSIs are also expected, to which at least the preprocessing capability will be provided.

As to image processors themselves, the local-parallel pipelined processor will be further improved to provide higher processing speeds. At the same time, the multiprocessor image processor will be applied in industry when the key processing element becomes more widely available. The image processor will become smaller and faster, and will have new functions, in response to the advancement of semiconductor technology, such as progress in system-on-chip configuration and wafer-scale integration. It may also be possible to realize 1-chip intelligent processors for high-level processing, and to combine these with 1-chip rather low-level image processors to achieve intelligent processing, such as knowledge-based or model-based processing. Based on these new developments, image processing and the resulting machine vision are expected to generate new values not merely for industry but also for all aspects of human life.

References

Abbreviations Used

ICPR: International (Joint) Conference on Pattern Recognition
IECE: Institute of Electronics and Communication Engineers of Japan
IEEE: Institute of Electrical and Electronics Engineers
IEEJ: Institute of Electrical Engineers of Japan
IEICE: Institute of Electronics, Information and Communication Engineers of Japan (formerly IECE)
IFAC: International Federation of Automatic Control
IJCAI: International Joint Conference on Artificial Intelligence
ISSCC: International Solid-State Circuit Conference
SICE: Society of Instrument and Control Engineers of Japan

1. M. Ejiri, *Proc. 2nd Int. Joint Conf. on Robotics Research* (Kyoto, 1984), pp. 166–172. Also in *Robotics Research* (The MIT Press, Cambridge, MA, 1985), pp. 461–467.
2. M. Ejiri, T. Uno, H. Yoda, T. Goto, and K. Takeyasu *Proc. 2nd IJCAI* (London, 1971), pp. 350–358.
3. M. Ejiri, T. Uno, H. Yoda, T. Goto, and K. Takeyasu, *IEEE Trans. Comput.* **C-21**, 2, 161–170 (1972).
4. M. Ohshima, *Electrotechnical Laboratory Report,* No. 826 (1982), in Japanese.
5. P. M. Will and K. S. Pennington, *Proc. 2nd IJCAI* (London, 1971), pp. 66–70.
6. K. Sato and S. Inokuchi, *1st Int. Conf. on Computer Vision* (London, 1987), pp. 657–661.
7. H. Yoda, S. Ikeda, and M. Ejiri, *Hitachi Rev.* **22**, 9, 362–365 (1973). Also in *Trans. SICE* **10**, 3, 284–289 (1974), in Japanese.
8. M. Ejiri, S. Kashioka, and H. Ueda, *Computers in Industry* **5**, 2, 107–113 (1984).
9. M. Ejiri, T. Uno, M. Mese, and S. Ikeda, *Computer Graphics and Image Proc.* **2**, 2/3, 326–339 (1973).

10. S. Kashioka, M. Ejiri, and Y. Sakamoto, *IEEE Trans. Syst. Man. Cybern.* **SMC-6,** 8, 562–570 (1976).
11. M. Mese, T. Miyatake, S. Kashioka, M. Ejiri, I. Yamazaki, and T. Hamada, *Proc. 5th IJCAI* (Cambridge, 1977), pp. 685–693.
12. T. Uno, M. Ejiri, and T. Tokunaga, *Pattern Recognition* **8,** 4, 201–208 (1976).
13. J.T. Olsztyn and R. Dewer, *Proc. 1st ICPR* (1973) pp. 503–513.
14. M. Baird, *Proc. 3rd ICPR* (1976).
15. S. Kashioka, M. Ejiri, and Y. Sakamoto, *Trans. IEEJ* **96-C,** 1, 9–16 (1976), in Japanese.
16. T. Fukushima, Y. Kobayashi, K. Hirasawa, T. Bandoh, M. Ejiri, and H. Kuwahara, *ISSCC Dig. Tech. Papers* 258–259 (1983).
17. T. Fukushima, Y. Kobayashi, K. Hirasawa, T. Bandoh, and M. Ejiri, *Trans. IECE* **J66-C,** 12, 959–966 (1983), in Japanese.
18. K. Kaneko, T. Nakagawa, A. Kiuchi, Y. Hagiwara, H. Ueda, H. Matsushima, T. Akazawa, T. Satoh, and J. Ishida, *ISSCC Dig. Tech. Papers* 158–159 (1987).
19. H. Yoda, Y. Ohuchi, Y. Taniguchi, and M. Ejiri, *IEEE Trans. Pattern Anal. and Machine Intell.* **PAMI-10,** 1, 4–16 (1988).
20. H. Yoda, J. Motoike, and M. Ejiri, *Proc. 4th IJCAI* (Tbilisi, 1975), pp. 620–627.
21. T. Uno, H. Matsushima, H. Yoda, K. Kitta, J. Motoike, T. Yasue, and T. Miyatake, *National Project Report on PIPS* (Pattern Information Processing System), pp. 127–135 (1980), in Japanese.
22. K. Okamoto, T. Hamada, and N. Akiyama, *Proc. 14th SICE Conf.,* No. 3606 (1975), in Japanese.
23. K. Haga, K. Nakamura, Y. Sano, N. Miyamori, and A. Komuro, *Fuji Jiho* **52,** 5, 294–298 (1979), in Japanese.
24. T. Miyatake, H. Matsushima, and M. Ejiri, *Proc. Nat. Conf. of IEICE,* No. 1554 (1987), in Japanese.
25. S. Kashioka, Y. Shima, and M. Ejiri, *Trans. IECE* **J68-D,** 5, 1103–1110 (1985), in Japanese.
26. T. Uno, S. Ikeda, H. Ueda, and M. Ejiri, in *Computer Vision and Sensor-based Robot,* G. G. Dodd and L. Rossol, eds. (Plenum, New York, 1979), pp. 101–116.
27. S. Nakahara, A. Maeda, and Y. Nomura, *Denshi Tokyo,* IEEE Tokyo Section **18,** 46–48 (1979).
28. Y. Nomura, S. Ito, and M. Naemura, *Mitsubishi Denki Giho* **53,** 12, 899–903 (1979), in Japanese.
29. S. Sawano, J. Ikeda, N. Utsumi, H. Kiba, Y. Ohtani, and A. Kikuchi,

Proc. Int. Conf. on Advanced Robotics (Tokyo, 1983), pp. 351–358. Also in *Robotica* (Cambridge University Press, 1984), pp. 41–46.

30. Y. Nakagawa, Y. Oshida, and T. Ninomiya, *Trans. SICE* **22**, 9, 982–987 (1986), in Japanese.

31. M. Mese, T. Uno, S. Ikeda, and M. Ejiri, *Trans. IEEJ* **94-C**, 5, 89–96 (1974), in Japanese.

32. N. Goto, K. Ichikawa, T. Kondo, and M. Kanemoto, *Toshiba Rev.* **33**, 6, 615–618 (1978), in Japanese.

33. Y. Shima, S. Kashioka, K. Kato, and M. Ejiri, *Trans. IECE* **J68-D**, 7, 1384–1391 (1985), in Japanese.

34. H. Sakou, H. Yoda, M. Ejiri, and O. Obata, *Trans. IEEJ* **107-C**, 8, 737–744 (1987), in Japanese.

35. Y. Nakagawa, H. Makihara, N. Akiyama, T. Numakura, and T. Nakagawa, *Proc. IFAC Symp. Information-Control Problems in Manufacturing Technology* (Tokyo, 1977), pp. 63–70.

36. Y. Hara, K. Okamoto, T. Hamada, and N. Akiyama, *Proc. 5th ICPR* (Miami, 1980), pp. 273–279.

37. Y. Hara, K. Okamoto, T. Hamada, and K. Nakagawa, *Trans. SICE* **19**, 11, 903–908 (1983), in Japanese.

38. T. Ninomiya and Y. Nakagawa, *IEEE Proc. Int. Workshop on Machine Vision and Machine Intelligence* (1987), pp. 346–351.

39. H. Sakou, H. Yoda, and M. Ejiri, *Trans. IECE* **J69-D**, 11, 1687–1696 (1986), in Japanese.

40. H. Yoda, H. Sakou, M. Ejiri, N. Hamamoto, H. Ishida, and K. Kudo, *1st Symp. on Image Sensing Technology for Industry*, No. 10-5 (1986), pp. 351–356, in Japanese.

41. H. Ueda, T. Uno, M. Ejiri, and K. Nakamura, *Proc. Nat. Conf. Inst. TV Engrs. of Japan* (1977), pp. 11–16, in Japanese.

42. S. Kashioka, T. Yasue, and Y. Shima, *Proc. Nat. Conf. of IECE*, No. 1266 (1980), in Japanese.

43. Y. Shima, S. Kashioka, and T. Yasue, *Trans. IECE* **J69-D**, 3, 417–426 (1986), in Japanese.

44. T. Nakano, O. Ozeki, and S. Yamamoto, *IECE Tech. Report* **85**, 147, PRL85-29 (1985), in Japanese.

45. M. Ejiri, S. Kakumoto, T. Miyatake, S. Shimada, and H. Matsushima, *Proc. 7th ICPR* (Montreal, 1984), pp. 1296–1305.

46. S. Kakumoto, T. Miyatake, S. Shimada, and M. Ejiri, *Trans. IECE* **J68-D**, 4, 829–836 (1985), in Japanese.

47. T. Asano, G. Kenwood, and S. Hata, *Trans. IECE* **J69-D**, 11, 1654–1661 (1986), in Japanese.

48. H. Kurokawa, T. Temma, K. Matsumoto, and M. Iwashita, *Proc. Nat. Conf. of IECE,* No. 1423 (1984).
49. T. Fukushima, Y. Kobayashi, S. Miura, and K. Asada, *Proc. 8th ICPR,* (Paris, 1986), pp. 38–41.
50. S. Kashioka, K. Kato, H. Ueda, M. Ejiri, and T. Noguchi, *Proc. 8th Annual British Robot Association Conf.* (Birmingham, 1985), pp. 123–132.
51. K. Kato, H. Ueda, S. Kashioka, H. Sakou, and O. Obata, *Hitachi Hyoron* **67,** 9, 731–734 (1985), in Japanese.
52. H. Ueda, K. Kato, and H. Matsushima, *Proc. Nat. Conf. of IECE,* No. 1634 (1986), in Japanese.
53. H. Ueda, K. Kato, H. Matsushima, K. Kaneko, and M. Ejiri, *Proc. Int. Conf. on Systolic Arrays* (San Diego, 1988), pp. 611–620.
54. M. Ejiri, *Proc. Joint Conf. of Electric- and Information-Related Societies,* No. 28–7 (1985), in Japanese.
55. M. Ejiri, *Industrial Image Processing* (Shokodo Co. Ltd., Tokyo, 1988), in Japanese.
56. Y. Shirai, in *The Psychology of Computer Vision,* P. H. Winston, ed. (McGraw-Hill, New York 1975).
57. M. Yachida and S. Tsuji, *Trans. IECE* **J59-D,** 3, 149–156 (1976), in Japanese.

Index

Japanese Technology Reviews